BRITANNICA
Mathematics in Context

Figuring All the Angles

NORTH POLE
welcomes you too

UTAH
Welcomes You

Britannica
ENCYCLOPÆDIA BRITANNICA EDUCATIONAL CORPORATION

Mathematics in Context is a comprehensive middle grades curriculum. It was developed in collaboration with the Wisconsin Center for Education Research, School of Education, University of Wisconsin–Madison and the Freudenthal Institute at the University of Utrecht, The Netherlands, with the support of National Science Foundation Grant No. 9054928.

National Science Foundation

Opinions expressed are those of the authors
and not necessarily those of the Foundation

ISBN 0-7826-1493-0
1 2 3 4 5 6 7 8 9 10   99 98 97

The *Mathematics in Context* Development Team

Mathematics in Context is a comprehensive middle grades curriculum. The National Science Foundation funded the National Center for Research in for Mathematical Sciences Education at the University of Wisconsin–Madison to develop and field-test the materials from 1991 through 1996. The Freudenthal Institute at the University of Utrecht in The Netherlands is the main subcontractor responsible for the development of the student assessment activities.

The initial version of *Figuring All the Angles* was developed by Jan de Lange, Martin van Reeuwijk, Els Feijs, and the staff of the Freudenthal Institute. It was adapted for use in American schools by James A. Middleton and Margaret A. Pligge.

National Center for Research in Mathematical Sciences Education Staff

Thomas A. Romberg
Director

Joan Daniels Pedro
Assistant to the Director

Gail Burrill
Coordinator
Field Test Materials

Margaret R. Meyer
Coordinator
Pilot Test Materials

Mary Ann Fix
Editorial Coordinator

Sherian Foster
Editorial Coordinator

James A. Middleton
Pilot Test Coordinator

Project Staff

Jonathan Brendefur
Laura J. Brinker
James Browne
Jack Burrill
Rose Byrd
Peter Christiansen
Barbara Clarke
Doug Clarke
Beth R. Cole

Fae Dremock
Jasmina Milinkovic
Margaret A. Pligge
Mary C. Shafer
Julia A. Shew
Aaron N. Simon
Marvin Smith
Stephanie Z. Smith
Mary S. Spence

Freudenthal Institute Staff

Jan de Lange
Director

Els Feijs
Coordinator

Martin van Reeuwijk
Coordinator

Project Staff

Mieke Abels
Nina Boswinkel
Frans van Galen
Koeno Gravemeijer
Marja van den Heuvel-Panhuizen
Jan Auke de Jong
Vincent Jonker
Ronald Keijzer

Martin Kindt
Jansie Niehaus
Nanda Querelle
Anton Roodhardt
Leen Streefland
Adri Treffers
Monica Wijers
Astrid de Wild

Acknowledgments

Several school districts used and evaluated one or more versions of the materials: Ames Community School District, Ames, Iowa; Parkway School District, Chesterfield, Missouri; Stoughton Area School District, Stoughton, Wisconsin; Madison Metropolitan School District, Madison, Wisconsin; Milwaukee Public Schools, Milwaukee, Wisconsin; and Dodgeville School District, Dodgeville, Wisconsin. Two sites were involved in staff development as well as formative evaluation of materials: Culver City, California, and Memphis, Tennessee. Two sites were developed through partnership with Encyclopædia Britannica Educational Corporation: Miami, Florida, and Puerto Rico. University Partnerships were developed with mathematics educators who worked with preservice teachers to familiarize them with the curriculum and to obtain their advice on the curriculum materials. The materials were also used at several other schools throughout the United States.

We at Encyclopædia Britannica Educational Corporation extend our thanks to all who had a part in making this program a success. Some of the participants instrumental in the program's development are as follows:

Allapattah Middle School
Miami, Florida
Nemtalla (Nikolai) Barakat

Ames Middle School
Ames, Iowa
Kathleen Coe
Judd Freeman
Gary W. Schnieder
Ronald H. Stromen
Lyn Terrill

Bellerive Elementary
Creve Coeur, Missouri
Judy Hetterscheidt
Donna Lohman
Gary Alan Nunn
Jakke Tchang

Brookline Public Schools
Brookline, Massachusetts
Rhonda K. Weinstein
Deborah Winkler

Cass Middle School
Milwaukee, Wisconsin
Tami Molenda
Kyle F. Witty

Central Middle School
Waukesha, Wisconsin
Nancy Reese

Craigmont Middle School
Memphis, Tennessee
Sharon G. Ritz
Mardest K. VanHooks

Crestwood Elementary
Madison, Wisconsin
Diane Hein
John Kalson

Culver City Middle School
Culver City, California
Marilyn Culbertson
Joel Evans
Joy Ellen Kitzmiller
Patricia R. O'Connor
Myrna Ann Perks, Ph.D.
David H. Sanchez
John Tobias
Kelley Wilcox

Cutler Ridge Middle School
Miami, Florida
Lorraine A. Valladares

Dodgeville Middle School
Dodgeville, Wisconsin
Jacqueline A. Kamps
Carol Wolf

Edwards Elementary
Ames, Iowa
Diana Schmidt

Fox Prairie Elementary
Stoughton, Wisconsin
Tony Hjelle

Grahamwood Elementary
Memphis, Tennessee
M. Lynn McGoff
Alberta Sullivan

Henry M. Flagler Elementary
Miami, Florida
Frances R. Harmon

Horning Middle School
Waukesha, Wisconsin
Connie J. Marose
Thomas F. Clark

Huegel Elementary
Madison, Wisconsin
Nancy Brill
Teri Hedges
Carol Murphy

Hutchison Middle School
Memphis, Tennessee
Maria M. Burke
Vicki Fisher
Nancy D. Robinson

Idlewild Elementary
Memphis, Tennessee
Linda Eller

Jefferson Elementary
Santa Ana, California
Lydia Romero-Cruz

Jefferson Middle School
Madison, Wisconsin
Jane A. Beebe
Catherine Buege
Linda Grimmer
John Grueneberg
Nancy Howard
Annette Porter
Stephen H. Sprague
Dan Takkunen
Michael J. Vena

Jesus Sanabria Cruz School
Yabucoa, Puerto Rico
Andreíta Santiago Serrano

John Muir Elementary School
Madison, Wisconsin
Julie D'Onofrio
Jane M. Allen-Jauch
Kent Wells

Kegonsa Elementary
Stoughton, Wisconsin
Mary Buchholz
Louisa Havlik
Joan Olsen
Dominic Weisse

Linwood Howe Elementary
Culver City, California
Sandra Checel
Ellen Thireos

Mitchell Elementary
Ames, Iowa
Henry Gray
Matt Ludwig

New School of Northern Virginia
Fairfax, Virginia
Denise Jones

Northwood Elementary
Ames, Iowa
Eleanor M. Thomas

Orchard Ridge Elementary
Madison, Wisconsin
Mary Paquette
Carrie Valentine

Parkway West Middle School
Chesterfield, Missouri
Elissa Aiken
Ann Brenner
Gail R. Smith

Ridgeway Elementary
Ridgeway, Wisconsin
Lois Powell
Florence M. Wasley

Roosevelt Elementary
Ames, Iowa
Linda A. Carver

Roosevelt Middle
Milwaukee, Wisconsin
Sandra Simmons

Ross Elementary
Creve Coeur, Missouri
Annette Isselhard
Sheldon B. Korklan
Victoria Linn
Kathy Stamer

St. Joseph's School
Dodgeville, Wisconsin
Rita Van Dyck
Sharon Wimer

St. Maarten Academy
St. Peters, St. Maarten, NA
Shareed Hussain

Sarah Scott Middle School
Milwaukee, Wisconsin
Kevin Haddon

Sawyer Elementary
Ames, Iowa
Karen Bush Hoiberg

Sennett Middle School
Madison, Wisconsin
Brenda Abitz
Lois Bell
Shawn M. Jacobs

Sholes Middle School
Milwaukee, Wisconsin
Chris Gardner
Ken Haddon

Stephens Elementary
Madison, Wisconsin
Katherine Hogan
Shirley M. Steinbach
Kathleen H. Vegter

Stoughton Middle School
Stoughton, Wisconsin
Sally Bertelson
Polly Goepfert
Jacqueline M. Harris
Penny Vodak

Toki Middle School
Madison, Wisconsin
Gail J. Anderson
Vicky Grice
Mary M. Ihlenfeldt
Steve Jernegan
Jim Leidel
Theresa Loehr
Maryann Stephenson
Barbara Takkunen
Carol Welsch

Trowbridge Elementary
Milwaukee, Wisconsin
Jacqueline A. Nowak

W. R. Thomas Middle School
Miami, Florida
Michael Paloger

Wooddale Elementary Middle School
Memphis, Tennessee
Velma Quinn Hodges
Jacqueline Marie Hunt

Yahara Elementary
Stoughton, Wisconsin
Mary Bennett
Kevin Wright

Site Coordinators

Mary L. Delagardelle—Ames Community Schools, Ames, Iowa

Dr. Hector Hirigoyen—Miami, Florida

Audrey Jackson—Parkway School District, Chesterfield, Missouri

Jorge M. López—Puerto Rico

Susan Militello—Memphis, Tennessee

Carol Pudlin—Culver City, California

Reviewers and Consultants

Michael N. Bleicher
Professor of Mathematics
University of Wisconsin–Madison
Madison, WI

Diane J. Briars
Mathematics Specialist
Pittsburgh Public Schools
Pittsburgh, PA

Donald Chambers
Director of Dissemination
University of Wisconsin–Madison
Madison, WI

Don W. Collins
Assistant Professor of Mathematics Education
Western Kentucky University
Bowling Green, KY

Joan Elder
Mathematics Consultant
Los Angeles Unified School District
Los Angeles, CA

Elizabeth Fennema
Professor of Curriculum and Instruction
University of Wisconsin-Madison
Madison, WI

Nancy N. Gates
University of Memphis
Memphis, TN

Jane Donnelly Gawronski
Superintendent
Escondido Union High School
Escondido, CA

M. Elizabeth Graue
Assistant Professor of Curriculum and Instruction
University of Wisconsin–Madison
Madison, WI

Jodean E. Grunow
Consultant
Wisconsin Department of Public Instruction
Madison, WI

John G. Harvey
Professor of Mathematics and Curriculum & Instruction
University of Wisconsin–Madison
Madison, WI

Simon Hellerstein
Professor of Mathematics
University of Wisconsin–Madison
Madison, WI

Elaine J. Hutchinson
Senior Lecturer
University of Wisconsin–Stevens Point
Stevens Point, WI

Richard A. Johnson
Professor of Statistics
University of Wisconsin–Madison
Madison, WI

James J. Kaput
Professor of Mathematics
University of Massachusetts–Dartmouth
Dartmouth, MA

Richard Lehrer
Professor of Educational Psychology
University of Wisconsin–Madison
Madison, WI

Richard Lesh
Professor of Mathematics
University of Massachusetts–Dartmouth
Dartmouth, MA

Mary M. Lindquist
Callaway Professor of Mathematics Education
Columbus College
Columbus, GA

Baudilio (Bob) Mora
Coordinator of Mathematics & Instructional Technology
Carrollton-Farmers Branch Independent School District
Carrollton, TX

Paul Trafton
Professor of Mathematics
University of Northern Iowa
Cedar Falls, IA

Norman L. Webb
Research Scientist
University of Wisconsin–Madison
Madison, WI

Paul H. Williams
Professor of Plant Pathology
University of Wisconsin–Madison
Madison, WI

Linda Dager Wilson
Assistant Professor
University of Delaware
Newark, DE

Robert L. Wilson
Professor of Mathematics
University of Wisconsin–Madison
Madison, WI

Dear Teacher,

Welcome! *Mathematics in Context* is designed to reflect the National Council of Teachers of Mathematics Standards for School Mathematics and to ground mathematical content in a variety of real-world contexts. Rather than relying on you to explain and demonstrate generalized definitions, rules, or algorithms, students investigate questions directly related to a particular context and construct mathematical understanding and meaning from that context.

The curriculum encompasses 10 units per grade level. *Figuring All the Angles* is designed to be the second in the geometry strand for grade 5/6, but it also lends itself to independent use—to help students learn about navigating in the real world using a compass and angular measurement.

In addition to the Teacher Guide and Student Books, *Mathematics in Context* offers the following components that will inform and support your teaching:

• *Teacher Resource and Implementation Guide*, which provides an overview of the complete system, including program implementation, philosophy, and rationale

• *Number Tools*, which is a series of blackline masters that serve as review sheets or practice pages involving number issues and basic skills

• *News in Numbers*, which is a set of additional activities that can be inserted between or within other units; it includes a number of measurement problems that require estimation.

• *Teacher Preparation Videos*, which present comprehensive overviews of the units to help with lesson preparation

Thank you for choosing *Mathematics in Context.* We wish you success and inspiration!

Sincerely,

The Mathematics in Context Development Team

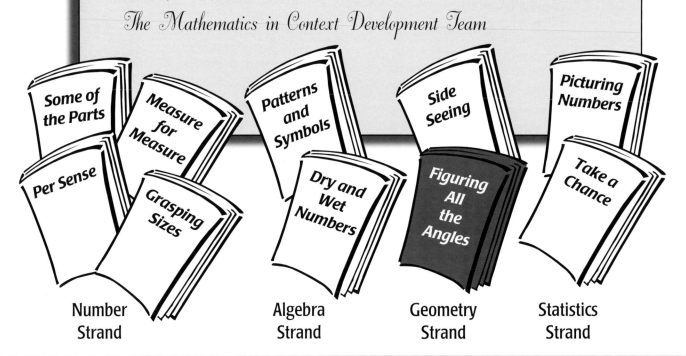

Number Strand   Algebra Strand   Geometry Strand   Statistics Strand

Overview

BRITANNICA

Mathematics in Context

How to Use This Book

This unit is one of 40 for the middle grades. Each unit can be used independently; however, the 40 units are designed to make up a complete, connected curriculum (10 units per level). There is a Student Book and a Teacher Guide for each unit.

Each Teacher Guide comprises elements that assist the teacher in the presentation of concepts and in understanding the general direction of the unit and the program as a whole. Becoming familiar with this structure will make using the units easier.

Each Teacher Guide consists of six basic parts:

- Overview
- Student Material and Teaching Notes
- Assessment Activities and Solutions
- Glossary
- Blackline Masters
- Try This! Solutions

Overview

Before beginning this unit, read the Overview in order to understand the purpose of the unit and to develop strategies for facilitating instruction. The Overview provides helpful information about the unit's focus, pacing, goals, and assessment, as well as explanations about how the unit fits in with the rest of the *Mathematics in Context* curriculum.

Note: After reading the Overview, view the Teacher Preparation Videotape that corresponds with the strand. The video models several activities from the strand.

Student Materials and Teaching Notes

This Teacher Guide contains all of the student pages (except the Try This! activities), each of which faces a page of solutions, samples of students' work, and hints and comments about how to facilitate instruction. Note: Solutions for the Try This! activities can be found at the back of the Teacher Guide.

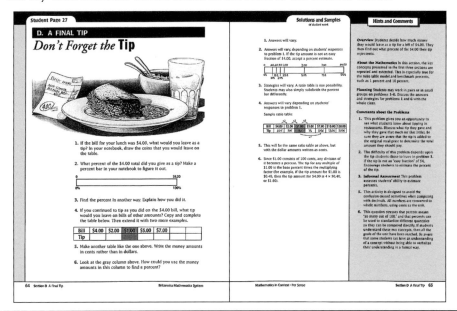

Each section within the unit begins with a two-page spread that describes the work students do, the goals of the section, new vocabulary, and materials needed, as well as providing information about the mathematics in the section and ideas for pacing, planning instruction, homework, and assessment.

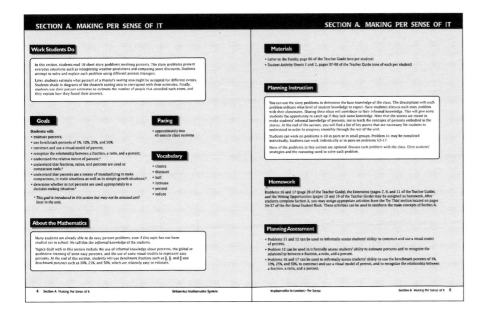

Assessment Activities and Solutions

Information about assessment can be found in several places in this Teacher Guide. General information about assessment is given in the Overview; informal assessment opportunities are identified on the teacher pages that face each student page; and the Assessment Activities section of this guide provides formal assessment opportunities.

Glossary

The Glossary defines all vocabulary words listed on the Section Opener pages. It includes mathematical terms that may be new to students, as well as words associated with the contexts introduced in the unit. (Note: The Student Book does not have a glossary. This allows students to construct their own definitions, based on their personal experiences with the unit activities.)

Blackline Masters

At the back of this Teacher Guide are blackline masters for photocopying. The blackline masters include a letter to families (to be sent home with students before beginning the unit), several student activity sheets, and assessment masters.

Try This! Solutions

Also included in the back of this Teacher Guide are the solutions to several Try This! activities—one related to each section of the unit—that can be used to reinforce the unit's main concepts. The Try This! activities are located in the back of the Student Book.

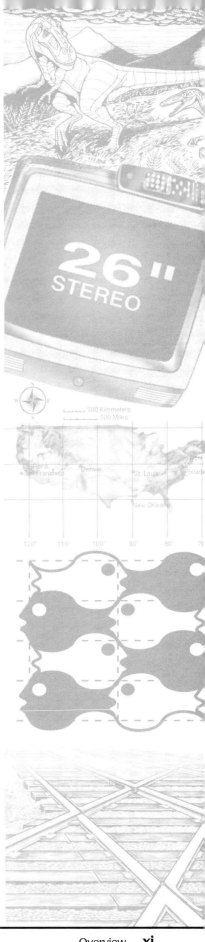

Unit Focus

In this unit, students use a compass and angular measurement to discover more about the world in which they live. The first activity focuses on the four cardinal directions (north, south, east, and west). Students explore the layouts of their classroom, their home, and their town in relation to these directions. As they progress through the unit, students investigate the relative nature of these directions in the context of names and locations of places in the United States. Using the notions of angle and distance, they find their way around a city and investigate airplane routes. Students describe trips in detail, making inferences about turns and resulting angles. In the last section, students investigate angles in mosaic tiles.

Mathematical Content

- using cardinal directions
- estimating directions, turns, and angles
- understanding the concepts of parallel and perpendicular
- estimating angle measurements
- measuring angles and distances
- using a rectangular or polar grid
- using vectors
- using scales, maps, and shapes

Prior Knowledge

This unit assumes that students have an understanding of the following:
- adding and subtracting three-digit numbers,
- the four cardinal directions—north, south, east, and west,
- left and right.

Planning and Preparation   Pacing: 14–20 days

Section	Work Students Do	Pacing*	Materials
A. A Sense of Direction	■ work with spatial orientation ■ use cardinal directions	1–2 days	■ Letter to the Family, Student Activity Sheets 1 and 2, magnetic compasses (one of each per student) ■ maps of your local area (one per pair or group) ■ copies of classroom floor plan, optional (one per student) ■ globe, optional (one per class) ■ atlases or maps, optional (four or five per class)
B. Finding Your Way	■ use a rectangular grid for orientation	2–3 days	■ Student Activity Sheets 3 and 4 (one of each per student) ■ string, optional (one piece per group) ■ copies of local maps and paper strips, optional (one per group)
C. Investigating North	■ investigate how flat maps can describe a sphere	1–2 days	■ Student Activity Sheet 5 (one per student) ■ orange; globe, optional; tennis ball, optional; transparency of the grid map, optional (one of each per class) ■ scissors (one pair per student) ■ tape (one roll per pair or group) ■ transparency of Student Activity Sheet 5, optional (one per class)
D. Directions	■ use headings and distances to set courses for travel	2–3 days	■ Student Activity Sheets 6 and 7, ruler (one of each per student) ■ Transparency Masters 1 and 2 (one large and one medium compass card per student) ■ U.S. maps (one per pair of students) ■ See page 51 of the Teacher Guide for a list of optional materials and quantities needed
E. Navigation and Orientation	■ use a polar grid for orientation	2–3 days	■ Student Activity Sheets 8–10 and compass cards, (one of each per student) ■ See page 65 of the Teacher Guide for a list of optional materials and quantities needed
F. Changing Directions: Turns	■ make turns to change direction ■ explore the relationship between a turn and its resulting angle	4–5 days	■ Student Activity Sheet 11, compass cards, rulers (one of each per student) ■ Transparency Master 3, optional (a few per class) ■ masking tape, (one roll per class) ■ meter sticks, optional (one per class) ■ grid paper, optional (one sheet per student)
G. Angles and Shapes	■ recognize angles in the world	2 days	■ See page 97 of the Teacher Guide for a list of optional materials and quantities needed

* One day is approximately equivalent to one 45-minute class session.

Preparation

In the *Teacher Resource and Implementation Guide* is an extensive description of the philosophy underlying both the content and the pedagogy of the *Mathematics in Context* curriculum. Suggestions for preparation are also given in the Hints and Comments columns of this Teacher Guide. You may want to consider the following:

• Work through the unit before teaching it. If possible, take on the role of the student and discuss your strategies with other teachers.

• Use the overhead projector for student demonstrations, particularly with overhead transparencies of the student activity sheets and any manipulatives used in the unit.

• Invite students to use drawings and examples to illustrate and clarify their answers.

• Allow students to work at different levels of sophistication. Some students may need concrete materials, while others can work at a more abstract level.

• Provide opportunities and support for students to share their strategies, which often differ. This allows students to take part in class discussions and introduces them to alternative ways to think about the mathematics in the unit.

• In some cases, it may be necessary to read the problems to students or to pair students to facilitate their understanding of the printed materials.

• A list of the materials needed for this unit is in the chart on page xiii.

• Try to follow the recommended pacing chart on page xiii. You can easily spend more time on this unit than the number of class periods indicated. Bear in mind, however, that many of the topics introduced in this unit will be revisited and covered more thoroughly in other *Mathematics in Context* units.

Resources

For Teachers	For Students
Books and Magazines *Mathematics Assessment: Myths, Models, Good Questions, and Practical Suggestions*, edited by Jean Kerr Stenmark (Reston, Virginia: The National Council of Teachers of Mathematics, Inc., 1991)	**Videos** Mathsphere Videos • *Rescue* • *Twin Peaks* (available from Encyclopædia Britannica)
Videos *Geometry Strand Teacher Preparation Video*	

Assessment

Planning Assessment

In keeping with the NCTM Assessment Standards, valid assessment should be based on evidence drawn from several sources. (See the full discussion of assessment philosophies in the *Teacher Resource and Implementation Guide*.) An assessment plan for this unit may draw from the following sources:

• Observations—look, listen, and record observable behavior.

• Interactive Responses—in a teacher-facilitated situation, note how students respond, clarify, revise, and extend their thinking.

• Products—look for the quality of thought evident in student projects, test answers, worksheet solutions, or writings.

These categories are not meant to be mutually exclusive. In fact, observation is a key part of assessing interactive responses and also key to understanding the end results of projects and writings.

Ongoing Assessment Opportunities

• Problems within Sections
To evaluate ongoing progress, *Mathematics in Context* identifies informal assessment opportunities and the goals that these particular problems assess throughout the Teacher Guide. There are also indications as to what you might expect from your students.

• Section Summary Questions
The summary questions at the end of each section are vehicles for informal assessment (see Teacher Guide pages 20, 34, 48, 62, 76, 94, and 108).

End-of-Unit Assessment Opportunities

In the back of this Teacher Guide, there are six short assessment activities and one thematic task, the Floriade Flower Exibition, that can be completed in two 45-minute class periods. For a more detailed description of the assessment activities, see the Assessment Overview (Teacher Guide pages 110 and 111).

You may also wish to design your own culminating project or let students create one that will tell you what they consider important in the unit. For more assessment ideas, refer to the charts on pages xvi and xvii.

Goals and Assessment

In the *Mathematics in Context* curriculum, unit goals, categorized according to cognitive procedures, relate to the strand goals and to the NCTM Curriculum and Evaluation Standards. Additional information about these goals is found in the *Teacher Resource and Implementation Guide.* The *Mathematics in Context* curriculum is designed to help students develop their abilities so that they can perform with understanding in each of the categories listed below. It is important to note that the attainment of goals in one category is not a prerequisite to attaining those in another category. In fact, students should progress simultaneously toward several goals in different categories.

	Goal	Ongoing Assessment Opportunities		End-of-Unit Assessment Opportunities
Conceptual and Procedural Knowledge	**1.** indicate a direction (heading) using cardinal directions and degrees	**Section A** **Section B** **Section C**	p. 8, #6 p. 20, #23 p. 30, #13 p. 46, #14c	East Wind Island, p. 152 I See You and You See Me, p. 152 The Cutting Machine, p. 153 Detecting a Fire, p. 154 Hide-and-Seek, p. 155
	2. identify a position using both rectangular and polar grids	**Section B** **Section E**	p. 26, #5 p. 28, #10 p. 30, #13 p. 68, #4	Detecting a Fire, p. 154 The Floriade Flower Exibition, pp. 157–161
	3. understand and use the relationships between turns and angles	**Section F**	p. 94, #22	The Floriade Flower Exibition, pp. 157–161
	4. estimate and measure distances on a map or grid, directions relative to north, turns and angles	**Section B** **Section D** **Section E** **Section F** **Section G**	p. 28, #8 p. 30, #13 p. 58, #10 p. 60, #11, #12 p. 72, #6 p. 88, #9 p. 106, #10	Hide-and-Seek, p. 155 Patchwork, p. 156 The Floriade Flower Exibition, pp. 157–161
	5. use the scale on a map to estimate distances	**Section A** **Section B** **Section D**	p. 8, #6 p. 28, #8 p. 60, #11, #12	Hide-and-Seek, p. 155 The Floriade Flower Exibition, pp. 157–161

	Goal	Ongoing Assessment Opportunities		End-of-Unit Assessment Opportunities
Reasoning, Communicating, Thinking, and Making Connections	**6.** compare rectangular and polar grid systems	**Section B** **Section E**	p. 34, #19 p. 72, #5	
	7. use directions, headings, turns, and angles to solve simple problems	**Section A** **Section D** **Section E** **Section F** **Section G**	p. 12, #13 p. 58, #10 p. 60, #11, #12 p. 72, #6 p. 74, #7 p. 86, #8 p. 100, #4	I See You and You See Me, p. 152 The Floriade Flower Exibition, pp. 157–161
	8. understand the relationships among turns, resulting angles, and the number of sides of a regular polygon	**Section F**	p. 90, #16	Patchwork, p. 156
	9. recognize that there are different ways to present information in a map and recognize the consequences of different presentation methods	**Section C**	p. 44, #11	The Floriade Flower Exibition, pp. 157–161

	Goal	Ongoing Assessment Opportunities		End-of-Unit Assessment Opportunities
Modeling, Nonroutine Problem-Solving, Critically Analyzing, and Generalizing	**10.** understand and use the relationships among directions, headings, turns, and angles	**Section D** **Section F**	p. 60, #11, #12 p. 86, #8 p. 94, #23	The Cutting Machine, p. 153 Detecting a Fire, p. 154 The Floriade Flower Exibition, pp. 157–161
	11. understand the relationship between the dynamic definition and the static interpretation of angles	**Section F**	p. 94, #23	Patchwork, p. 156
	12. use directions, turns, and angles in combination with scales, distances, and the implicit use of vectors to solve more complex problems	**Section D**	p. 60, #11, #12	The Floriade Flower Exibition, pp. 157–161

More about Assessment

Scoring and Analyzing Assessment Responses

Students may respond to assessment questions with various levels of complexity, mathematical sophistication, and elaboration. Each student's response should be considered for the mathematics that it shows, and not judged on whether or not it includes an expected response. Responses to some of the assessment questions may be viewed as either correct or incorrect, but many answers will need flexible judgment by the teacher. Descriptive judgments related to specific goals and partial credit often provide more helpful feedback than percentage scores.

Openly communicate your expectations to all students, and report achievement and progress for each student relative to those expectations. When scoring students' responses, try to think about how they are progressing toward the goals of the unit and the strand.

Student Portfolios

Generally, a portfolio is a collection of student-selected pieces that is representative of a student's work. A portfolio may include evaluative comments by you or by the student. See the *Teacher Resource and Implementation Guide* for more ideas on portfolio focus and use.

A comprehensive discussion about the contents, management, and evaluation of portfolios can be found in *Mathematics Assessment: Myths, Models, Good Questions, and Practical Suggestions*, pp. 35–48.

Student Self-Evaluation

Self-evaluation encourages students to reflect on their progress in learning mathematical concepts, their developing abilities to use mathematics, and their dispositions toward mathematics. The following examples illustrate ways to incorporate student self-evaluations as one component of your assessment plan.

- Ask students to comment, in writing, on each piece they have chosen for their portfolios and on the progress they see in the pieces overall.

- Give a writing assignment entitled "What I Know Now about [a math concept] and What I Think about It." This will give you information about each student's disposition toward mathematics as well as his or her knowledge.

- Interview individuals or small groups to elicit what they have learned, what they think is important, and why.

Suggestions for self-inventories can be found in *Mathematics Assessment: Myths, Models, Good Questions, and Practical Suggestions*, pp. 55–58.

Summary Discussion

Discuss specific lessons and activities in the unit—what the student learned from them and what the activities have in common. This can be done in whole-class discussion, in small groups, or in personal interviews.

Connections across the *Mathematics in Context* Curriculum

Figuring All the Angles is the second unit in the geometry strand. The map below shows the complete *Mathematics in Context* curriculum for grade 5/6. It indicates where the unit fits in the number strand and where it fits in the overall picture.

A detailed description of the units, the strands, and the connections in the *Mathematics in Context* curriculum can be found in the *Teacher Resource and Implementation Guide*.

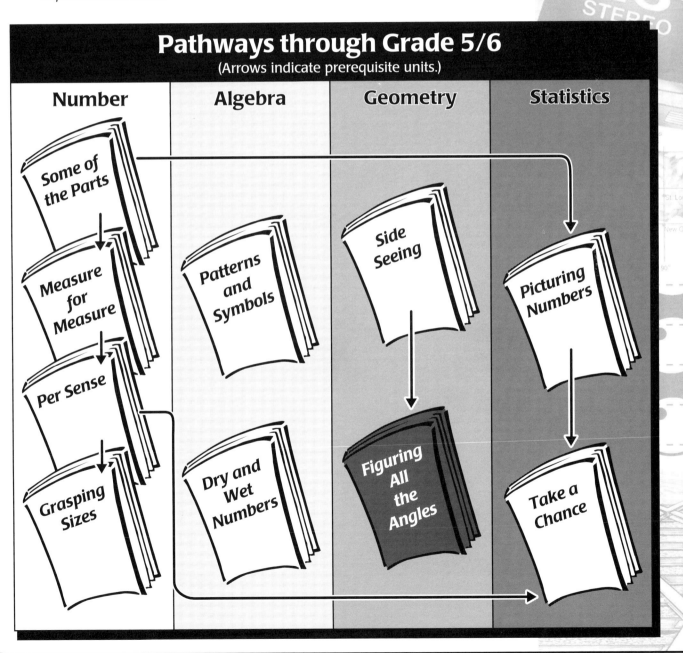

Pathways through Grade 5/6
(Arrows indicate prerequisite units.)

Number	Algebra	Geometry	Statistics
Some of the Parts			
Measure for Measure	Patterns and Symbols	Side Seeing	Picturing Numbers
Per Sense			
Grasping Sizes	Dry and Wet Numbers	Figuring All the Angles	Take a Chance

Grade 5/6

Side Seeing

Figuring All the Angles

Grade 6/7

Reallotment

Made to Measure

Grade 7/8

Packages and Polygons

Ways to Go

Triangles and Beyond

Looking at an Angle

Grade 8/9

Triangles and Patchwork

Going the Distance

Digging Numbers

Connections within the Geometry Strand

On the left is a map of the geometry strand; this unit, *Figuring All the Angles,* is highlighted.

Hans Freudenthal defined geometry as "grasping space, . . . the space that the child must learn to know, explore, conquer, in order to live, breathe, and move better in it." This notion of "grasping space" is one of the primary components of the *Mathematics in Context* geometry strand. *Side Seeing,* the first unit in the geometry strand, deals primarily with this notion.

Figuring All the Angles is the second unit in the geometry strand. Geometry is presented through the concept of grasping space to help students represent and make sense of the world. Orienteering and navigation together are one of three substrands of geometry. Initially, students explore the four cardinal directions. A change of perspective to include relative directions leads to a refinement of the four cardinal directions to include southeast, northwest, southwest, and northeast. Through the context of navigation, students further refine direction measurement by using degrees and distance. They work with both rectangular and polar grid systems. An informal approach to using vectors within a given context leads to more formal notions of angle measurement.

Exploration of space is revisited in the units *Looking at an Angle, Triangles and Beyond,* and *Going the Distance.*

The Geometry Strand

Grade 5/6

Side Seeing
Exploring the relationship between three-dimensional shapes and drawings of them, seeing from different points of view, and building structures from drawings.

Figuring All the Angles
Estimating and measuring angles and investigating direction, vectors, and rectangular and polar coordinates.

Grade 7/8

Packages and Polygons
Recognizing geometric shapes in real objects and representations, constructing models, and investigating properties of regular and semi-regular polyhedra.

Looking at an Angle
Recognizing vision lines in two and three dimensions; identifying and drawing shadows and blind spots; identifying the isomorphism of vision lines, light rays, flight paths, and so forth; understanding the relationship between angles and the tangent ratio; and computing with the tangent ratio.

Ways to Go
Reading and interpreting different kinds of maps, comparing different types of distances, progressing from one two-dimensional model to another (from a diagram to a map to a photograph to a graph), and drawing graphs and networks. (*Ways to Go* is also in the statistics strand.)

Triangles and Beyond
Exploring the interrelationships of the sides and angles of triangles as well as the properties of parallel lines and quadrilaterals, constructing triangles, and using transformations to become familiar with the concepts of congruence and similarity.

Grade 6/7

Reallotment
Measuring regular and irregular areas; discovering links between area, perimeter, surface area, and volume; and using English and metric units.

Made to Measure
Measuring length (including circumference), volume, and surface area using metric units.

Grade 8/9

Triangles and Patchwork
Understanding similarity and using it to find unknown measurements for similar triangles and developing the concept of ratio through tessellation.

Going the Distance
Using the Pythagorean theorem to investigate distances, scales, and vectors and using slope, tangent, area, square root, and contour lines.

Digging Numbers
Using the properties of height, diameter, and radius to determine whether or not various irregular shapes are similar; predicting length using graphs and formulas; exploring the relationship between three-dimensional shapes and drawings of them; and using length-to-width ratios to classify various objects. (*Digging Numbers* is also in the statistics strand.)

Connections with Other *Mathematics in Context* Units

Helping students comprehend space is an integral part of the geometry strand. *Side Seeing* begins this exploration, followed by *Figuring All the Angles*. *Looking at an Angle* continues the orientation and navigation theme begun in *Figuring All the Angles*.

The following mathematical topics that are included in the unit *Figuring All the Angles* are introduced or further developed in other *Mathematics in Context* units.

Topics Revisited in Other Units

Topic	Unit	Grade
direction	*Patterns and Symbols**	5/6
	*Dry and Wet Numbers**	5/6
	*Graphing Equations**	8/9
distance	*Looking at an Angle*	7/8
	*Ways to Go****	7/8
	Triangles and Patchwork	8/9
	Going the Distance	8/9
rectangular grids	*Patterns and Symbols**	5/6
	*Dry and Wet Numbers**	5/6
	*Operations**	6/7
circular grids	*Graphing Equations**	8/9
	Going the Distance	8/9
shapes	*Reallotment*	6/7
	Packages and Polygons	7/8
	Triangles and Beyond	7/8
angles	*Made to Measure*	6/7
	Packages and Polygons	7/8
	Looking at an Angle	7/8
	Triangles and Beyond	7/8
	Triangles and Patchwork	8/9
	Going the Distance	8/9
vectors	*Graphing Equations**	8/9
	Going the Distance	8/9
maps	*Patterns and Symbols**	5/6
	*Operations**	6/7
	*Insights into Data****	7/8
	Going the Distance	8/9

   * These units in the algebra strand also help students make connections to ideas about geometry.
  ** These units in the geometry strand also help students make connections to ideas about geometry.
 *** These units in the statistics strand also help students make connections to ideas about geometry.

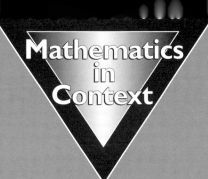

Student
Materials
and Teaching
Notes

Student Book
Table of Contents

Dear Student,

Welcome to *Figuring All the Angles.*

In this unit, you will learn to locate objects relative to their own positions and describe cities using directions such as north, south, east, and west; and headings such as 0° north and 180° south.

You will learn to describe the locations of planes using a model of a radar screen.

By the end of this unit, you should understand something about using directions and turns to describe locations. You will be introduced to ideas that will help you solve geometry problems later in your study of mathematics.

Sincerely,

The Mathematics in Context Development Team

Work Students Do

In this section, students explore the world in which they live. In the first activity, students investigate the relativity of the four cardinal directions, particularly the direction north. Students also examine local maps to determine in which directions different towns lie, relative to the location of their own town. They discuss reasons for some of the names of regions and states, such as the Midwest, the Southwest, South Dakota, and West Virginia. Students also work with the more refined directions northwest, southwest, northeast, and southeast. Finally, they explore how shadows resulting from the Sun can provide clues about direction and time.

Goals

Students will:

- indicate a direction using cardinal directions;
- estimate directions relative to north;*
- use the scale on a map to estimate distances;
- use directions to solve simple problems.

  * This goal is introduced in this section and assessed in other sections of the unit.

Pacing

- approximately one to two 45-minute class sessions

Vocabulary

• direction	• relative
• east	• south
• north	• southeast
• northeast	• southwest
• northwest	• west

About the Mathematics

Spatial orientation is the ability to describe location and movement in the world. Direction is an important part of spatial orientation. Initially, the four cardinal directions are used to describe locations. The eight wind directions describe locations in more detail. Natural indicators of direction, such as the position of the Sun, are also important for spatial orientation.

Materials

- Letter to the Family, page 134 of the Teacher Guide (one per student)

- Student Activity Sheets 1 and 2, pages 135 and 136 of the Teacher Guide (one of each per student)

- maps or copies of maps of your local area, pages 7 and 9 of the Teacher Guide (one per pair or group of students)

- magnetic compasses, pages 7 and 9 of the Teacher Guide (one per student)

- copies of classroom floor plan, page 7 of Teacher Guide, optional (one per student)

- globe, pages 13 and 15 of the Teacher Guide, optional (one per class)

- atlases or maps, pages 15 and 21 of the Teacher Guide, optional (four or five per class)

Planning Instruction

It is important for students to be able to use directions to describe locations. To introduce this section, ask students to write down directions to the cafeteria or to the school library. Encourage them to share their directions aloud. Ask students about the orientation of the classroom: *Does the door face north? What directions do the windows face?*

Students can work on problem 1 as a whole class and on problems 2–23 in pairs or in small groups.

Problem 19 is optional. If time is a concern, this problem may be skipped.

Homework

Problem 8 (page 8 of the Teacher Guide), problems 15–17 (page 14 of the Teacher Guide), problems 20 and 21 (page 18 of the Teacher Guide), and the Writing Opportunity (page 21 of the Teacher Guide) may be assigned as homework. After students complete Section A, you may assign appropriate activities from the Try This! section, located on pages 47–50 of the *Figuring All the Angles* Student Book. The Try This! activities reinforce the key mathematical concepts introduced in this section.

Planning Assessment

- Problem 6 can be used to informally assess students' ability to use the scale on a map to estimate distances and to indicate a direction using cardinal directions.

- Problem 13 can be used to informally assess students' ability to use directions to solve simple problems.

- Problem 23 can be used to informally assess students' ability to indicate a direction using cardinal directions.

A. A SENSE OF DIRECTION

Getting Your Sense of Direction

Activity

For the following activity you will need a magnetic compass and a map of your town or city.

1. Point towards **north.** Is everyone in your class pointing in the same **direction?**

2. **a.** Sketch your classroom. Draw an arrow pointing north on each of the desks.

   b. Do all the arrows point in the same direction?

   c. Will lines from the arrows ever meet?

3. Where is **south?**

4. The position of the Sun in the sky is related to the direction south. At what time of day is the Sun in the southern part of the sky?

5. In which direction from the classroom is your school's playground?

1. Answers will vary, but all students should point in the same general direction.

2. **a.** Sketches will vary, depending on your classroom and on the details that students provide in their sketches. An example of a student sketch is shown below.

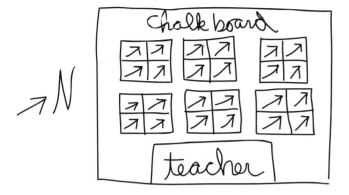

   b. Yes, all the arrows should point in the same direction.

   c. Yes, the lines from the arrows are almost parallel. Eventually, they meet at the North Pole.

3. South is opposite north; pointing south means pointing to the South Pole.

4. In the northern hemisphere, the Sun is in the southern part of the sky at noon and travels from east to west. The position of the Sun also depends on the time of year and the latitude.

5. Answers will vary. Use a compass to verify the direction of the playground.

Materials maps or copies of maps of your local area (one per pair or group); magnetic compasses (one per student); copies of classroom floor plan, optional (one per student)

Overview Students explore their notions of directions, particularly the direction north.

About the Mathematics Most students will have had some experience with the four cardinal directions—north, south, east, and west. This first activity builds an understanding of relative directions. Children often have difficulty pointing north, and they do not realize that all arms should be (almost) parallel when everyone points north.

Planning You may decide to start this activity by asking the class to work on problem **1.** Then students can work on problems **2–5** in pairs or in small groups. Discuss students' answers to these problems.

Comments about the Problems

1. You can ask students to first point toward north without looking at their classmates' arms. Students may point in completely different directions. Then have them observe and discuss the different directions in which the arms point. Ask students: *How do you know or how can you determine which direction is north in our classroom?* [Students may suggest using magnetic compasses, the Sun's position, and so on.]

   Invite students to point toward north again. Mark the wall indicating the spot where one of the students is pointing. Ask students: *When pointing north, should all of you point to this mark?* [No, students seated east of the student whose direction is marked should point more to the east, and students to the west should point more to the west.]

2. **a.** See if students understand where north is. If they are having difficulty, you may want to ask them to use a magnetic compass to find north and then position rulers on their desks pointing in that direction.

4. If possible, allow students to go outside and determine the direction of the Sun before, at, and after noon. Actually observing the Sun's positions may help students solve problems involving the Sun and shadows.

Use a map of your local area to answer problems **6** and **7**.

6. Name a town about 50 miles away and point in the direction of that town. Describe this direction.

7. If you traveled north from your school, which towns would you pass through?

Use a compass to answer problems **8** and **9**.

8. Sketch the room where you sleep. Be sure to include windows on your sketch. Point out where north is on the sketch.

9. Compare the sketches of all the students in your class. How many of the windows in the sketches face south?

6. Answers will vary, depending on your local area. All cities mentioned should be on a circle that has a 50-mile radius with your town at the center.

7. Answers will vary based on your local map.

8. Sketches will vary. One example of a student sketch is shown below.

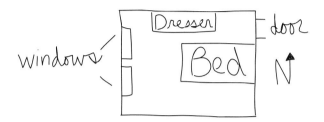

9. Answers will vary. In the sketch above, no windows are facing south.

Materials maps or copies of maps of your local town (one per pair or group); magnetic compasses (one per student)

Overview Students use a local map to explore the directions of different towns relative to the location of their own town. They also sketch their bedrooms, label the rooms' windows, and identify north.

Planning Students can work on problems **6–9** in pairs or in small groups. Problem **6** can be used for assessment, and problem **8** may be assigned as homework. After students complete problem **9,** a final class discussion will reinforce their understanding of directions.

Comments about the Problems

6. Informal Assessment This problem assesses students' ability to use the scale on a map to estimate distances and to indicate a direction using cardinal directions. Some students may use more detailed directions than north, south, east, and west.

8. Homework This problem may be assigned as homework. Have students share their sketches with the class.

North, East, South, West

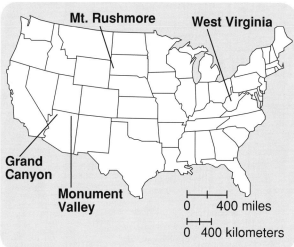

Use the partial map of the United States on **Student Activity Sheet 1** to answer these questions.

Monument Valley, in Arizona, has many spires and mesas that did not erode as fast as the land around them. You may have seen the picture on the right in Western movies. Western movies often describe how the West was settled.

10. Why is the West called *the West?*

On one side of Mount Rushmore, the heads of four United States presidents have been carved out of the mountain. Mount Rushmore is located in South Dakota.

11. Is South Dakota in the South? Why is it called *South Dakota?*

10. In the 1800s, west was the direction in which the people from the East were moving. Also, the West is close to the westernmost edge of North America.

11. South Dakota lies in the north of the United States, but south of North Dakota. Thus, it is called South Dakota because of its location relative to North Dakota.

Materials Student Activity Sheet 1 (one per student)

Overview Students explore reasons for the directional names of locations and states, such as South Dakota.

About the Mathematics These problems involve the relative nature of cardinal directions. A person's position determines the direction in which he or she needs to travel in order to arrive at a destination; for example, some people may travel south, while others travel north to reach the same location.

Planning Students can work in pairs or in small groups on problems **10** and **11.**

Comments about the Problems

10–11. Invite students to discuss the relative nature of directions.

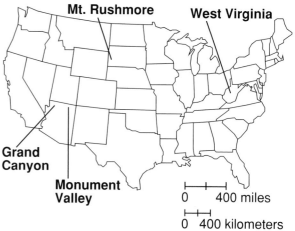

Mt. Rushmore

West Virginia

Grand Canyon

Monument Valley

0   400 miles

0   400 kilometers

The Grand Canyon is located in the area of the United States called the *Desert Southwest.*

12. Is the Grand Canyon in the ***southwest*** of the United States?

13. In what direction do people from San Francisco travel to visit the Grand Canyon? Draw a route from San Francisco to the Grand Canyon on **Student Activity Sheet 1.**

You have probably read about the North Pole.

14. Explain why the word *North* in North Pole has a different meaning than *West* in West Virginia. (West Virginia is labeled on **Student Activity Sheet 1.**)

12. Yes, geographically speaking, the Grand Canyon lies in the southwest of the United States.

13. To travel to the Grand Canyon, people from San Francisco have to travel to the east, but also to the south—hence southeast. Students' drawings will vary. Sample drawing shown below.

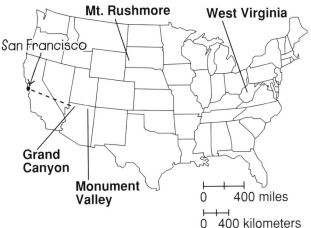

14. *North* in North Pole is different in that the North Pole is in the north for everybody. It is a fixed location (and is not the same place as the magnetic North Pole, which is at the north end of Earth's magnetic field). West Virginia uses the word *west* to describe its location in relation to the state of Virginia. Everyone calls the state "West Virginia," but it is not west of every location.

Materials Student Activity Sheet 1 (one per student); globe, optional (one per class)

Overview Students use the terms *southwest* and *southeast* to describe the location of the Grand Canyon. They also explore the difference between the meanings of *North* in North Pole and *West* in West Virginia.

Planning Students can work on problems **12–14** in pairs or in small groups. You may use problem **13** for assessment.

Comments about the Problems

13. Informal Assessment Problem **13** assesses students' ability to use directions to solve simple problems. San Francisco is not labeled on the U.S. map on Student Activity Sheet 1. Students need to know or look up the location of San Francisco to answer the question.

14. You may want to provide a globe on which students can locate West Virginia and the North Pole. To help students understand the North Pole's fixed or absolute location, you can ask, *What is north of the North Pole?* [nothing] Encourage students to locate their state and find a location to the west of the state. Then ask them to identify locations farther west until they end up in their home state again. This activity illustrates the relative nature of east and west (for example, there is no East or West Pole).

Did You Know? Magnetic compasses have needles that are magnetized. Because they move freely, the needles take on the same orientation as the north–south magnetic field of Earth. This magnetic field is not quite parallel to the north–south axis of the globe, but it is close enough to make a compass a reasonably good guide.

Source: Encyclopædia Britannica Micropædia.

Chicago is a city in the Midwest.

15. Do you think *midwest* is a good description for the position of Chicago on the map of the United States? Explain.

16. How would you describe the position of Chicago on the United States map?

17. Think of more locations—inside or outside the United States—with names that relate to compass directions.

15. Answers will vary. Some students may say no because Chicago doesn't lie in the "middle of the West" geographically speaking. Others may say yes because the United States is west of other locations (such as Europe) and because Chicago is about in the "middle" of the U.S.

16. "A little northeast of the middle" would be a typical description.

17. Answers will vary. Some examples with names related to compass directions include: North and South Korea, the North Sea, South Carolina, South America, and South Africa.

Materials globe, optional (one per class); atlases or maps, optional (four or five per class)

Overview Students describe the position of Chicago on a U.S. map. They also identify other locations with names related to compass directions.

Planning Students can work on problems **15–17** in pairs or in small groups. You may wish to have a brief class discussion about students' answers to problems **15** and **16**. Problems **15–17** can be assigned as homework.

Comments about the Problems

15–16. Homework These problems may be assigned as homework. Make sure students correctly locate Chicago on the map. Have them share their answers with the class.

17. Homework This problem may also be assigned as homework. A globe, atlas, or map may be helpful if students are unfamiliar with world geography.

Interdisciplinary Connection Problem **15** provides a historical connection. In the United States, the definition of the West changed dramatically as people migrated farther westward. Not long ago, nearly the entire population of the United States lived in the area between the Atlantic Coast and Chicago.

Photo: Courtesy of Wisconsin Department of Tourism.

The state capitol of Wisconsin is located in Madison. The capitol building is unusual because of the four identical wings radiating from the huge central dome.

The four wings of the building point in the four compass directions: north, south, **east,** and **west.** The south wing is indicated in the drawing below.

18. Label the directions of the remaining three wings on **Student Activity Sheet 2.**

Look at the shadow of the dome in the photo.

19. a. From which direction is the Sun shining?

    b. Around what time of day was the picture taken?

SOUTH

18. See picture below.

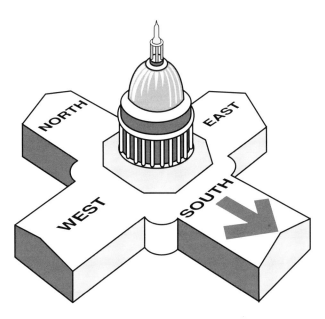

19. a. The Sun is shining from the southeast (because the shadow is on the north inside corner of the west wing).

   b. The picture was taken in the morning because the Sun is in the east.

Materials Student Activity Sheet 2 (one per student)

Overview Students label the directions in which a building's wings point. Based on a dome's shadow, they also determine the Sun's position.

About the Mathematics The position of the Sun is a natural indicator of direction. Students should have a global understanding of the relationship between the position of the Sun and the time of day. Although most students are familiar with the Sun and its shadows, they are not always aware of the relationship between them. It is important that students get the opportunity to investigate these phenomena, but at this stage it is not necessary that they understand everything about the solar system.

Planning Problem **19** is optional. Students can work on this problem in pairs or in small groups. If you decide to do this problem, you may want to have a class discussion before students begin to work on it.

Comments about the Problems

19. Not all students may know that they have to look eastward to see a sunrise, and look westward to see a sunset. If students (while working on problem **4**) did not fully explore the connection between the Sun's position and the time of day, you may want to encourage them to observe playground shadows during recess. Students can also compare morning shadows to afternoon shadows.

To help students find a strategy, you may want to ask one or more of the following questions: *In what direction do you have to look to see the Sun at noon? Where would you find the shadow of the dome in the picture at noon?* [If you are anywhere in the contiguous United States, the Sun will be south at noon. The shadow of the dome would point toward north at noon.]

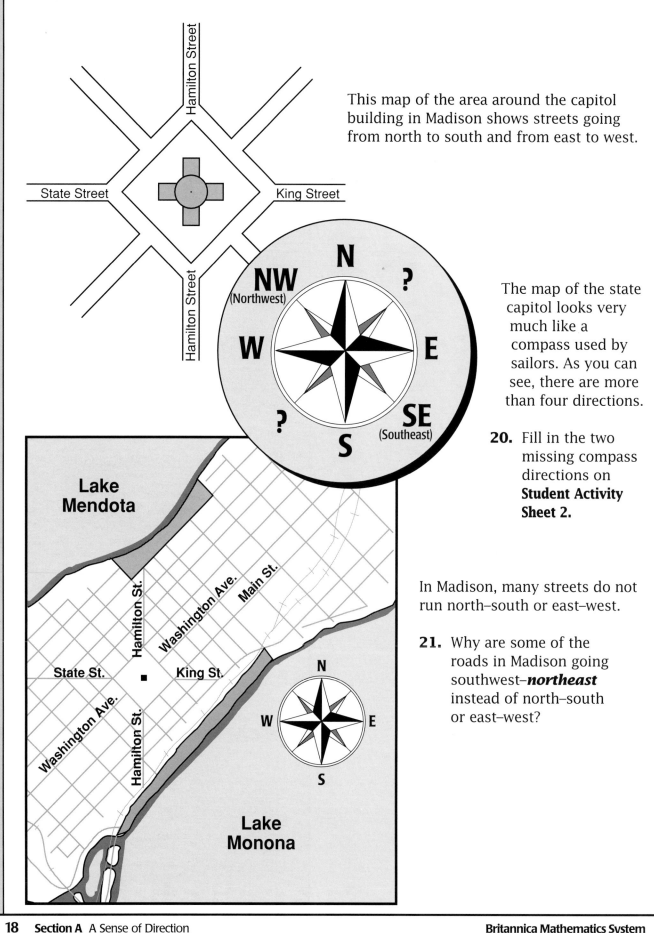

This map of the area around the capitol building in Madison shows streets going from north to south and from east to west.

The map of the state capitol looks very much like a compass used by sailors. As you can see, there are more than four directions.

20. Fill in the two missing compass directions on **Student Activity Sheet 2.**

In Madison, many streets do not run north–south or east–west.

21. Why are some of the roads in Madison going southwest–**northeast** instead of north–south or east–west?

20. northeast and southwest

21. Because the area of land that makes up Madison runs roughly southwest–northeast, many streets are angled in the same direction to make best use of the land space.

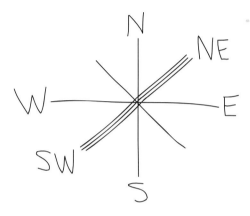

Materials Student Activity Sheet 2 (one per student)

Overview Students investigate the compass directions southwest and northeast. They determine why some of the streets in Madison, Wisconsin, run southwest–northeast.

About the Mathematics This page can be connected to the radial or polar grid, which will be developed further in Section E.

Planning Students can work on problems **20** and **21** in pairs or in small groups. You may decide to assign these problems as homework.

Comments about the Problems

20. Homework This problem may be assigned as homework. Ask students to think of names for the directions between north and west, south and west, and so on. Some students may know the term *south–southeast* (SSE) for the direction between south and southeast.

The picture below shows more compass headings.

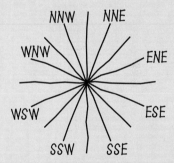

21. Homework This problem may be assigned for homework. Have students share their answers with the class.

Summary

North, south, east, and west are directions you can use to find places on a map and in the real world. You can combine them to be more specific; for example, you can say northeast, **northwest, southeast,** and southwest.

Directions are **relative.** For example, South Dakota is not in the South, but it is south of North Dakota. West Virginia is in the eastern United States, but it is west of Virginia.

Summary Questions

The North Pole is a unique place on Earth.

22. Why is it unique? Name another unique place on Earth?

23. How would you describe the positions of Hawaii and Alaska *relative* to the United States mainland?

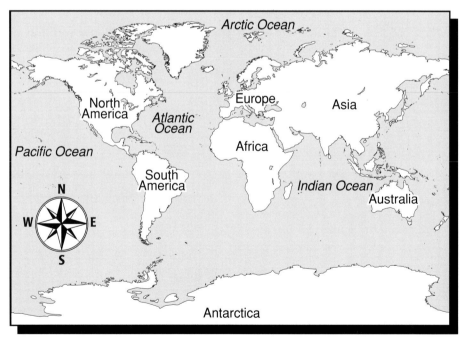

© 1996, Encyclopædia Britannica, Inc.

22. The North Pole is unique because of its fixed position: it is north of every location. Another such unique place is the South Pole.

23. Hawaii is located southwest of the mainland. Alaska is located northwest of the contiguous United States.

Materials atlases or maps, optional (four or five per class)

Overview Students read the Summary. They discuss unique locations, such as the North Pole. They also describe the positions of Hawaii and Alaska relative to the U.S. mainland.

Planning Students can work on problems **22** and **23** in pairs or in small groups. You can use problem **23** for assessment. After students complete Section A, you may assign appropriate activities from the Try This! section, located on pages 47–50 of the Student Book, as homework.

Comments about the Problems

22. The geographical North Pole has a fixed location. However, the magnetic North Pole does not. For this problem, we're concerned with the geographical North Pole.

23. Informal Assessment Problem **23** assesses students' ability to indicate a direction using cardinal directions. Make sure that students can locate Hawaii and Alaska on a U.S. map or atlas.

Materials atlases or maps, optional (four or five per class)

Writing Opportunity Have students write a paragraph in their journals about the cardinal directions.

Work Students Do

Students investigate rectangular grids as they explore the town of Sunray, where streets run east–west and avenues run north–south. Students locate houses and shops and determine distances between different buildings in Sunray. Given directions and distances, students mark the locations of different buildings on a map. They explore two types of distances: as-the-crow-flies and taxicab, or grid, distances. Students also examine and use a map of Provo, Utah, to give directions to different downtown locations. They draw a map of an imaginary town with eight roads leaving the city's center in the eight wind directions. Students then compare this map to the map of Sunray.

Goals

Students will:

• indicate a direction using cardinal directions;

• identify a position using a rectangular grid;

• use the scale on a map to estimate distances;

• use directions to solve simple problems;*

• compare rectangular and polar grid systems;

• estimate and measure distances on a map or grid and directions relative to north.

* This goal is introduced in this section and assessed in other sections of the unit.

Pacing

• approximately two to three 45-minute class sessions

Vocabulary

• as-the-crow-flies distance

• headwind

• locate

• parallel

• perpendicular

• taxicab distance

About the Mathematics

This section highlights two methods for computing distances. Distance can be calculated as a straight line between two points (*as the crow flies*). It can also be calculated along a pathway. If the path is on a rectangular grid, the distance is called a *taxicab distance*. On a rectangular grid, two fixed axes are used and a point is located by its distance from both axes.

Materials

- Student Activity Sheets 3 and 4, pages 137 and 138 of the Teacher Guide (one of each per student)

- strings for measuring as-the-crow-flies distances, page 29 of the Teacher Guide, optional (one piece per pair or group of students)

- copies of local maps, page 31 of the Teacher Guide, optional (one per pair or group of students)

- paper strips, page 33 of the Teacher Guide, optional (one per pair or group of students)

Planning Instruction

Introduce this section by having students study a local map. Ask them the following questions about the streets on the map: *In what directions do the streets run? Are they parallel or perpendicular? What are some of their names? Do you notice any patterns in the way the streets are named? What is the distance between [name two locations] on the map?*

Students can work on problems 1–5, 7–9, and 10–18 in pairs or in small groups. Problems 6, 19, and 20 should be completed individually.

Their are no optional problems. Discuss each problem with the class. Elicit students' strategies.

Homework

Problems 7–9 (page 28 of the Teacher Guide) and 19 and 20 (page 34 of the Teacher Guide) may be assigned as homework. The Writing Opportunity and Extension (page 35 of the Teacher Guide) may also be assigned as homework. After students complete Section B, you may assign appropriate activities from the Try This! section, located on pages 47–50 of the *Figuring All the Angles* Student Book. The Try This! activities reinforce key mathematical concepts introduced in this section.

Planning Assessment

- Problems 5 and 10 can be used to informally assess students' ability to identify a position using a rectangular grid.

- Problem 8 can be used to informally assess students' ability to use the scale on a map to estimate distances.

- Problem 13 can be used to informally assess students' ability to indicate a direction using cardinal directions, to identify a position using a rectangular grid, and to estimate and measure distances on a map or grid and directions relative to north.

- Problem 19 can be used to informally assess students' ability to compare rectangular and polar grid systems.

Sunray (1850)

MAIN STREET

Small towns often develop along a single road. In Sunray, the first road was called Main Street. Like many towns, Sunray developed in the mid-1800s, when the population of the United States was shifting from the East to the West.

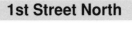

1st Street North

Town Hall

Main Street

1st Street South

Lincoln Avenue

Soon the city expanded and needed more streets. New streets were built either **parallel** or **perpendicular** to Main Street. The roads perpendicular to Main Street were called avenues.

1. Draw a map of Sunray with three streets north of Main Street, two streets south of Main Street, two avenues west of Lincoln Avenue, and three avenues east of Lincoln Avenue.

2. **a.** Why is it convenient to have the avenues run north–south when the streets run east–west?

   b. How are the streets named in the town where you go to school?

1. See diagram below.

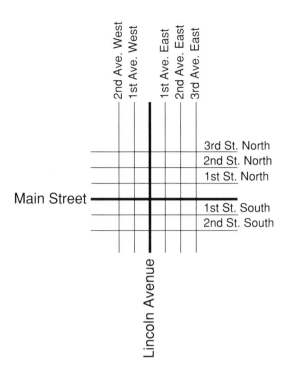

2nd Ave. West
1st Ave. West
1st Ave. East
2nd Ave. East
3rd Ave. East

3rd St. North
2nd St. North
1st St. North

Main Street

1st St. South
2nd St. South

Lincoln Avenue

2. a. Answers will vary. Students may say that this provides easy access (and often the shortest path) to many locations. And the convenient names provide additional information about specific locations.

b. Answers will vary depending on your local area.

Overview By drawing streets parallel to Main Street and avenues perpendicular to Main Street, students extend the map of a small, but expanding, town.

About the Mathematics There are different ways to locate places. The problems in the previous section were an introduction to the concepts underlying the polar coordinate system (in which directions and distances are used to locate places). The polar coordinate system will be further investigated in Section E.

In this section, students develop and use the rectangular grid, and the concepts of parallel and perpendicular lines play an important role. Later, this will lead to the use of a rectangular coordinate system in which two distances are needed to locate places.

Planning Students can work on problems **1** and **2** in pairs or in small groups. Encourage students to share their answers to problem **2.**

Comments about the Problems

1. Notice how students interpret *perpendicular* and *parallel*. The goal here is to develop an informal sense of these terms. For example, you might ask students whether certain streets are parallel, perpendicular, or neither.

2. b. Have students think about the system according to which the streets in their neighborhood are named. If your local area is similar to the example, you may want to have students return to the map of Madison on Student Book page 7 for a different arrangement of streets. Again, the concepts of parallel and perpendicular can be discussed.

SUNRAY (1900)

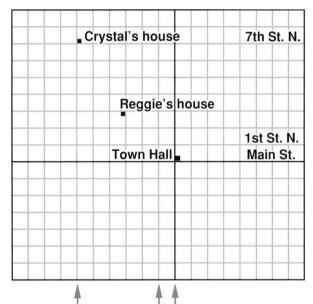

Crystal's house   7th St. N.

Reggie's house

1st St. N.

Town Hall   Main St.

6th Ave. W.   1st Ave. W.   Lincoln Ave.

Sunray has grown. It has many streets and avenues. The plan of the city looks like a grid—a combination of horizontal and vertical lines.

Crystal lives at the corner of 7th Street North and 6th Avenue West. Reggie lives at the corner of 3rd Street North and 3rd Avenue West.

Use the copy of this grid on **Student Activity Sheet 3** to help you answer the problems below.

3. Crystal and Reggie are friends. If Crystal wants to visit Reggie at his house, how far will she have to walk?

Crystal and Reggie want to meet each other for lunch. They like both Tony's Tortellini on the corner of Main Street and 7th Avenue West and Ella's Deli on 5th Street North and Lincoln Avenue.

4. Which restaurant is closer to their homes?

A theater is going to be built three blocks south of Main Street. The contractor has a choice of different building sites, all between 3rd Avenue East and 5th Avenue West.

5. a. How far from Crystal's house could the theater be built?

  b. Which location would be closest to Crystal's house?

3. Seven blocks. Students can count 4 blocks south and 3 blocks east or 3 blocks east and 4 blocks south, or any other combination.

4. Ella's Deli is closer to their homes. It is 8 blocks from Crystal's house and 5 blocks from Reggie's house, for a total of 13 blocks.

Tony's Tortellini is 8 blocks from Crystal's house and 7 blocks from Reggie's house, for a total of 15 blocks.

5. a. The answer depends on the choice of theater locations. The theater can be built between 3rd Avenue West and 3rd Avenue East on 3rd Street South. The farthest possible site is 19 blocks from Crystal's house, at 3rd Street South and 3rd Avenue East.

b. The closest location is 11 blocks from Crystal's house at 3rd Street South and 5th Avenue West.

Materials Student Activity Sheet 3 (one per student)

Overview Students locate places at intersections of streets and avenues. They also solve problems about distances on a rectangular grid.

About the Mathematics Students may remember the Pizza Delivery activity from the unit *Patterns and Symbols,* in which they had to find different routes on a rectangular grid. The problems on this page emphasize the concepts of location and distance. The following page explicitly describes two different ways to determine distances.

Planning Students can work in pairs or in groups on problems **3–5.** Problem **5** can be used for assessment.

Comments about the Problems

3. Some students may have trouble deciding where the houses are on the blocks. If students are having difficulty, tell them to imagine that the houses are located right at the intersection of the street and the avenue, and to start counting blocks from the intersection.

4–5. Encourage students to explain their reasoning.

5. Informal Assessment This problem assesses students' ability to identify a position using a rectangular grid.

Sunray (1995)

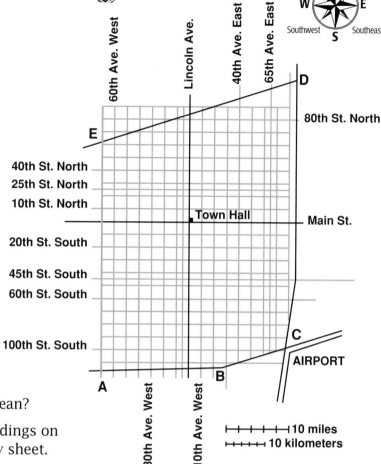

Here is a map of Sunray in 1995. Use the copy of this map on **Student Activity Sheet 4** to solve the following problems:

The baseball stadium lies exactly 7 miles south of Town Hall.

6. Draw the stadium on the map.

The basketball coliseum is 10 miles northwest of Town Hall.

The amusement park is 12 miles southeast of Town Hall.

The shopping mall is 3 miles southwest of intersection E.

All distances are **as the crow flies.**

7. a. What does *as the crow flies* mean?

   b. Mark the positions of the buildings on the large map on your activity sheet.

8. How far do you have to drive to go from the shopping mall to Town Hall?

Distances that follow the lines of a map grid are sometimes called **taxicab distances.**

9. Explain why distances on a grid are called *taxicab distances.*

10. Make a list of at least three buildings that you might want to build in Sunray (for example, schools, malls, museums). Give directions to your neighbor and see if he or she can **locate** the positions of your buildings on the map. Use any information available except street names.

A skyscraper with a restaurant on top will be built in Sunray. The height of the tower will be 300 feet. Because the city does not want any accidents, the tower has to be built at least 10 miles from the airport.

11. On the map, outline the area where the tower cannot be built.

6. See map below. The stadium should be somewhere around 35th Street South and Lincoln Avenue.

7. a. It means in a straight line from point to point, as if you could fly and not have to use streets.

   b. See map below.

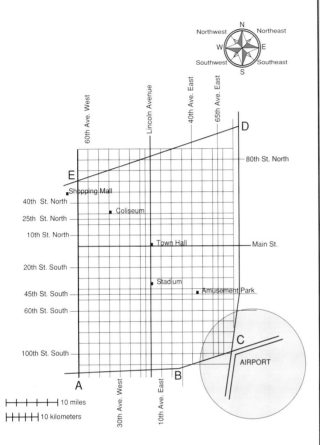

8. The mall is about 26 miles away from Town Hall by road.

9. Answers will vary. Sample response:

   Taxis have to drive on the streets and cannot take shortcuts.

10. Answers will vary. One example would be a school located 10 miles north and 10 miles east of Town Hall (located at 50th St. North and 50th Ave. East).

11. See the gray area on the map above.

Materials Student Activity Sheet 4 (one per student); strings for measuring as-the-crow-flies distances, optional (one piece per group)

Overview Given directions and distances, students draw the locations of different buildings on a map. They explore two different types of distances: *as the crow flies* and *taxicab* distances.

About the Mathematics Distance can be calculated as a straight line between two points, called *as the crow flies*. A distance can also be calculated along a path on a grid, called a *taxicab* distance.

Planning Students can work individually on problem **6.** They can work in pairs or in small groups on problems **7–9** and **11.** Problem **8** can be used for individual or group assessment. Problem **10** can be used for an assessment and must be done in pairs.

Comments about the Problems

6–10. The shift in scale may be difficult for students to see. You can emphasize the differences by having students contrast this map with the one on Student Book page 10. Encourage students to measure with a strip of paper or string rather than with a ruler.

6. You might ask an introductory question: *How do you find seven miles?* [Three blocks equal 6 miles, and half a block is 1 mile.]

7. Homework This problem may be assigned as homework. Note that the shopping mall is outside the grid system.

8. Informal Assessment This problem assesses students' ability to use the scale on a map to estimate distances. Students have to use the information from problem **6.** The shopping mall is 3 miles southwest of intersection E. Traveling from the shopping mall to the Town Hall, a driver could travel eastward and southward.

9. Homework This problem may be assigned as homework. Have students share their explanations with the class.

10. Informal Assessment This problem assesses students' ability to identify a position using a rectangular grid, directions, and distances. Students' directions should include a fixed reference point such as the Town Hall, a number of blocks, and a direction.

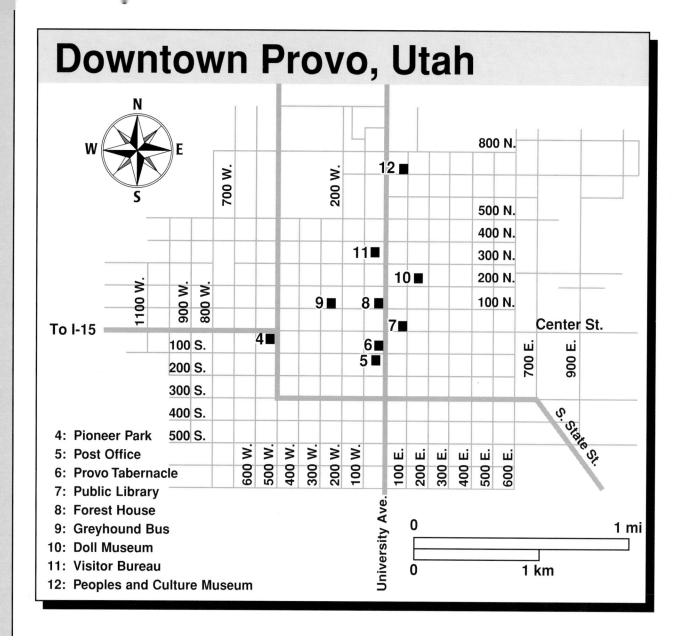

Downtown Provo, Utah

Key:
4: Pioneer Park
5: Post Office
6: Provo Tabernacle
7: Public Library
8: Forest House
9: Greyhound Bus
10: Doll Museum
11: Visitor Bureau
12: Peoples and Culture Museum

12. Look at the map above. How are the systems of naming streets alike for Sunray and Provo? How are they different?

The Greyhound Bus Station is number 9 on the map.

A passenger has just arrived at the bus station. He needs directions to the Visitor Bureau (number 11).

13. Give him directions.

At the Peoples and Culture Museum (number 12), someone tells a tourist that Pioneer Park is about a half hour's walk directly southwest of the museum.

14. Do you agree? Why or why not?

12. Answers may vary. Sample response:

The systems are nearly the same except that Provo's grid is much finer (there are more streets in a mile). Another difference is that *S. State St.* goes off at an angle in Provo.

13. Answers will vary. One possible set of directions:

Travel east from the bus station to 200 W. and turn left (north). Go 2 blocks north to the corner of 200 W. and 300 N. Turn right (east). Look for signs identifying the *Visitor Bureau.*

14. Answers may vary. Pioneer Park is not *exactly* southwest of the museum. Each block is about 0.1 mile. An average walking speed is 3 miles per hour. It is 13 blocks (in taxicab distance) or 1.3 miles from the Culture Museum to Pioneer Park (about a half hour's walk).

Materials copies of local maps, optional (one per group)

Overview Students use a map of Provo, Utah, to give directions around its downtown area.

About the Mathematics Distance can be expressed with different units, such as metric or standard units. But sometimes a distance is expressed in time. In problem **14,** distance is expressed in time (a half hour's walk) rather than miles or blocks. Students may have seen road signs that express distances as driving times. Not only in this unit, but also in other units such as *Grasping Sizes*, students develop reference points (for instance, in one hour you can walk about three miles, or about four kilometers) to make connections between these different units of distance.

Planning Students can work in pairs or in small groups on problems **12–14.** You may decide to have a class discussion about students' answers to problem **12.** Problem **13** can be used for assessment. You may want to use a local map instead of (or in addition to) the Provo map.

Comments about the Problems

13. Informal Assessment This problem assesses students' ability to indicate a direction using cardinal directions; to identify a position using a rectangular grid; and to estimate and measure distances on a map or grid and directions relative to north.

14. Students should notice that the problem gives two pieces of information, direction and distance. You may want students to discuss how fast an average person walks and explore how their answers differ due to their perceptions of walking speed. You may also have students discuss why a taxicab distance is appropriate for this problem.

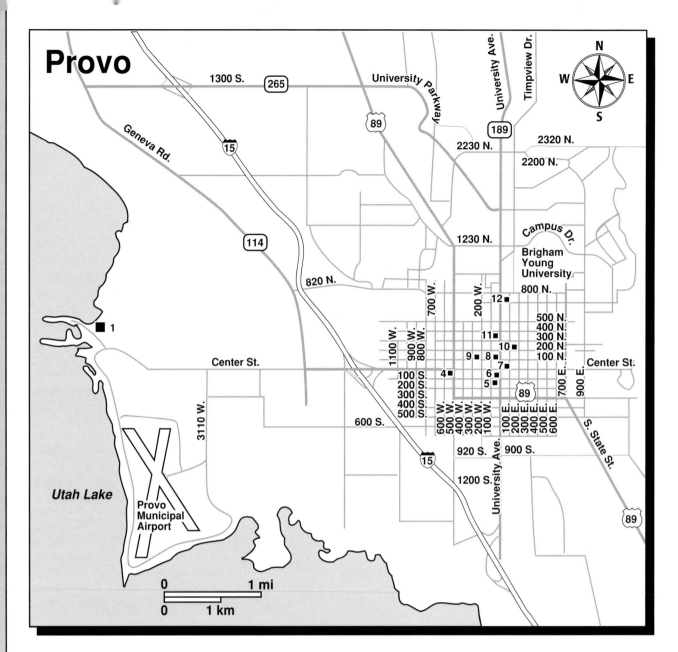

Provo

1300 S. 265

University Parkway

University Ave.

Timpview Dr.

N
W E
S

89

189

Geneva Rd.

15

2230 N.

2320 N.

2200 N.

114

1230 N.

Campus Dr.

Brigham
Young
University

820 N.

800 N.

700 W.

200 W.

12

1100 W.

900 W.

800 W.

500 N.
400 N.
300 N.
200 N.
100 N.

11

10

9 8

7

Center St.

Center St.

4

6
5

100 S.
200 S.
300 S.
400 S.
500 S.

89

700 E.

900 E.

3110 W.

600 S.

600 W.
500 W.
400 W.
300 W.
200 W.
100 W.

100 E.
200 E.
300 E.
400 E.
500 E.
600 E.

S. State St.

15

920 S. 900 S.

University Ave.

89

Utah Lake

Provo
Municipal
Airport

1200 S.

0   1 mi

0   1 km

15. In what directions does Interstate 15 run?

16. How many miles west and how many miles south of downtown Provo, where highways 89 and 189 intersect, is the airport?

Runways of airports are usually built in such a way that planes land with a **headwind.**

17. In what direction do you think the wind usually blows in Provo?

A developer wants to build a lookout tower in Utah Lake State Park, at a point $\frac{1}{4}$ mile west of the square labeled 1 on the map.

18. Can you think of a reason why she will not get a permit?

15. Roughly speaking, Interstate 15 runs northwest–southeast (with a nearly north–south jog in the middle).

16. About 3 miles west (look at 3110 West Avenue on the map) and 1 mile south.

17. Answers may vary. Sample response:

The wind usually blows from a southerly direction (ranging from southeast to southwest) or from a northerly direction (ranging from northeast to northwest).

18. Answers may vary. Sample response:

The tower will be in the direction of the northwest runway and too close.

Materials paper strips, optional (one per group)

Overview Students use a map of Provo and its surrounding area to determine distances and directions.

About the Mathematics A discussion of headwind in problem **17** foreshadows the airplane context used later in Section D.

A *headwind* is a wind blowing in the direction opposite that in which the plane is flying.

Planning Students can work on problems **15–18** in small groups. Encourage students to share their reasoning with each other.

Comments about the Problems

15. Notice whether students discover that on U.S. maps even-numbered highways run east–west and odd-numbered highways run north–south. If students have worked on the unit *Patterns and Symbols*, they should be able to distinguish between even and odd numbers.

16. Students can estimate using a paper strip and the scale on the map.

17. Some students may not understand the term *headwind*. To help them, you may want to ask the following question: *Why would a pilot like to land with a headwind?* [A headwind blows directly against the course of an aircraft, so a headwind helps slow the plane during landing.] Some students may not realize that a northerly wind blows southward. Wind directions are always named for the direction from which the wind blows.

Summary

Sometimes streets and avenues in a city are named so that residents and visitors can find places easily. Some cities have east–west streets and north–south avenues. For example, 5th Avenue West and 3rd Street North identify the exact location on a map or in a city. Distances can be measured as *taxicab distances* or *as the crow flies*.

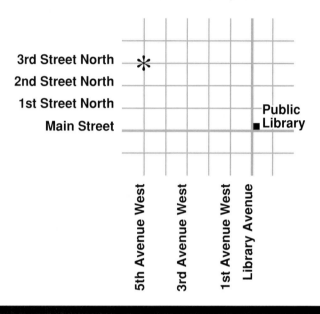

Summary Questions

19. a. Draw a map of a town that has Town Hall in the center and eight main roads leading from the center in the directions N, S, E, W, NW, NE, SW, and SE.

b. How is the map of Sunray different from the town you just drew?

20. Would *taxicab distances* on a street–avenue system be the same as distances measured *as the crow flies*? Explain.

19. a. Drawings will vary. One possibility is shown below.

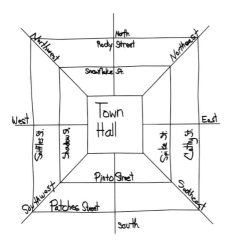

b. Answers may vary. Possible answer:

In Sunray, all of the blocks are square because there are no streets in the directions NW, NE, SW, SE, and it is easy to find a place if you know its address. In the town I drew, the streets are named after my pets and my friends' pets.

20. Answers may vary. Sample response:

Taxicab distances are usually longer than as-the-crow-flies distances. The distances could be the same if you have to go in one direction and a road goes in that direction.

Overview Students read the Summary. They draw a map of a town with eight roads leading from the center in eight different directions. They then compare this map to the map of Sunray. Finally, students compare *taxicab distances* to distances measured *as the crow flies.*

Planning Problem **19** can be used for assessment. Problems **19** and **20** can be assigned as homework. After students complete Section A, you may assign appropriate activities from the Try This! section, located on pages 47–50 of the Student Book, as homework.

Comments about the Problems

19. Informal Assessment This problem assesses students' ability to compare rectangular and polar grid systems. This problem can also be assigned as homework. Students are challenged to create a map that differs from the maps of Sunray and Provo. The map of Sunray is a rectangular grid, while the map students create may look like the polar grid they will investigate in Section E. Encourage informal comparisons rather than complicated answers involving terms such as *polar grid* or *rectangular grid.*

20. Homework This problem may be assigned as homework. Have students share their answers with the class.

Writing Opportunity Ask students to write their answers to problem **20** in their journals.

Extension As an extension to problem **19,** ask students to design their own city and create a brochure for the city's chamber of commerce to give to visitors.

Work Students Do

In this section, students investigate how flat maps can be used to describe a sphere. They explore the distortion of flat maps by attaching a paper map to an orange and then flattening a curved section of orange peel. Students begin to understand the difficulty in accurately representing larger regions (states and countries) on a flat grid. Students also work with the concepts of *latitude* and *longitude,* although these terms are not formally used in the section.

Goals

Students will:

- indicate a direction using cardinal directions;

- identify a position using both rectangular and polar grids;*

- recognize that there are different ways to present information in a map and the consequences of different presentation methods;

- estimate directions relative to north.*

  This goal is assessed in other sections of the unit.

Pacing

- approximately one to two 45-minute class sessions

Vocabulary

- rectangular grid
- sphere

About the Mathematics

The use of flat models, or maps, distorts directions and distances on curved surfaces, resulting in inaccurate representations of spherical objects, such as Earth. Although lines of longitude meet at Earth's poles, longitude lines are often drawn as parallels on a flat map. Longitude and latitude lines often form a rectangular grid system on a map. This section also explores the importance of the North Pole as a geographical reference point.

Materials

- Student Activity Sheet 5, page 139 of the Teacher Guide (one per student)
- globe, pages 41, 43, and 47 of the Teacher Guide, optional (one per class)
- transparency of the grid map, page 41 of the Teacher Guide, optional (one per class)
- orange, page 43 of the Teacher Guide (one per class)
- tennis ball, page 43 of the Teacher Guide, optional (one per class)
- scissors, page 47 of the Teacher Guide (one pair per student)
- tape, page 47 of the Teacher Guide (one roll per pair or group of students)
- transparency of Student Activity Sheet 5, page 47 of the Teacher Guide, optional (one per class)

Planning Instruction

You may want to introduce this section with a class discussion about the photographs of Earth taken from the *Apollo X* spacecraft. Encourage students to discuss their ideas. You may want to ask: *Many years ago, what did people think about the shape of Earth?* [People once thought that Earth had the shape of a flat disk and that sailors would fall off if they sailed too far in one direction.]

Student can work individually on problems 1 and 15–16, and they can solve problems 2–14 in pairs or in small groups.

Problem 8 is optional. If time is a concern, this problem may be omitted or assigned as homework.

Homework

Any problems begun in class can be assigned as homework. Problem 16 (page 48 of the Teacher Guide) offers a writing opportunity that students can complete at home. The Writing Opportunities (pages 43 and 45 of the Teacher Guide) can also be assigned as homework. After students complete Section C, you may assign appropriate activities from the Try This! section, located on pages 47–50 of the *Figuring All the Angles* Student Book. The Try This! activities reinforce the key mathematical concepts introduced in this section.

Planning Assessment

- Problem 11 can be used to informally assess students' recognition of the different ways to present information in a map and the consequences of different presentation methods.
- Problem 14c can be used to informally assess students' ability to indicate a direction using cardinal directions.

C. INVESTIGATING NORTH

Up North

Photo: Courtesy of National Aeronautics and Space Administration.

The Apollo X spacecraft was flying just 70 miles from the Moon when it took this picture of Earth. The photograph clearly shows that Earth is a **sphere.**

On the left and below are pictures you would see if you moved toward Earth from far away.

1. Describe the location of the United States.

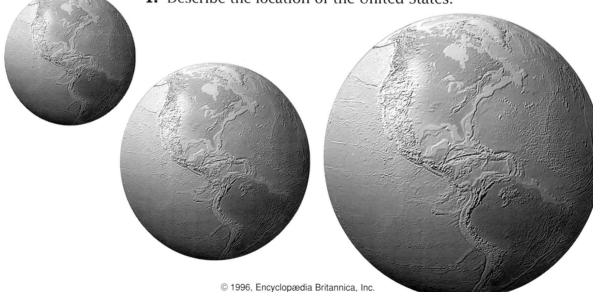

1. Answers will vary. Sample response:

The Unites States is located toward the center of the picture of Earth, in the top half (northern hemisphere).

Overview Students examine pictures of Earth, one of which was taken from the *Apollo X* spacecraft, and describe the location of the United States.

About the Mathematics A map is a projection, a systematic method of transferring the features of Earth's surface onto a flat sheet of paper. Different projections maintain or distort different properties, such as area, distance, and direction.

Planning You may want to start this section with a class discussion about the pictures on Student Book page 15. Students can work individually on problem **1.**

Comments about the Problems

1. This question introduces the spherical nature of Earth. Students may have learned the term *northern hemisphere* (the upper half of Earth) in geography class.

Did You Know? One of the most famous projections was invented in 1569 by Gerardus Mercator. He used a cylindrical projection, which displays parallels as horizontal lines and meridians as vertical lines (see fig. 1 below). Despite the great distortions in higher latitudes, the Mercator projection has many advantages. These maps have traditionally been used for navigational charts because compass bearings may be plotted as straight segments. The only navigational charts that are not Mercator projections are great-circle charts and charts of the polar regions (see fig. 2 below).

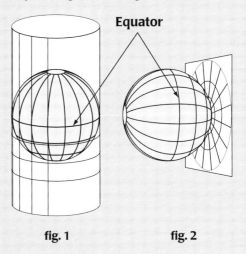

Equator

fig. 1   fig. 2

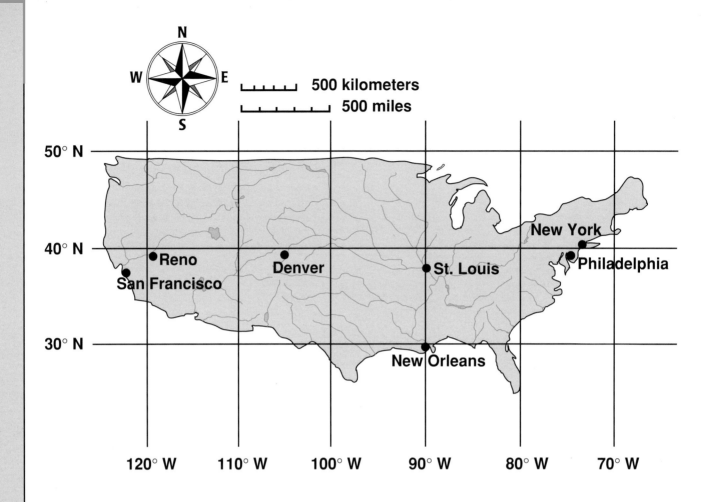

500 kilometers

500 miles

Above is a grid map of the United States. The location of St. Louis, Missouri, can be described as 38°N and 90°W.

2. Describe the locations of a) Reno, Nevada; b) Denver, Colorado; c) New Orleans, Louisiana; and d) Philadelphia, Pennsylvania.

Two planes take off at the same time, one from Denver and one from Philadelphia. They fly due north.

3. Could the planes ever meet? Explain your answer.

Two planes take off at the same time, one from St. Louis and one from New Orleans. They fly due west.

4. Could the planes ever meet? Explain your answer.

2. a. Reno, Nevada: 39°N, 119°W

   b. Denver, Colorado: 39°N, 105°W

   c. New Orleans, Louisiana: 30°N, 90°W

   d. Philadelphia, Pennsylvania: 39°N, 75°W

3. Yes, the two planes could meet at the North Pole (if they travel at the same speed).

4. No, the planes will never meet. They are flying parallel to each other.

Materials transparency of the grid map, optional (one per class); globe, optional (one)

Overview Students examine and use a grid map to locate U.S. cities. They also determine whether planes flying in the same direction will ever meet.

About the Mathematics In the previous sections, students investigated and used flat maps to locate places using distances and directions. Lines of latitude and longitude on a flat map are similar to a rectangular grid system. Lines of longitude on a sphere, however, meet at the poles.

Planning You may want to connect this page to the previous section (in which students identified buildings located at street intersections) by displaying an overhead transparency of the grid map. Ask students to describe the cities' locations by referring to the latitude and longitude lines as streets. Students can work in pairs or in small groups on problems **3** and **4** to facilitate discussion about the problems.

Comments about the Problems

2. This problem emphasizes identifying locations on a map using a grid system. It is similar to those in Section B, which involved a grid system of streets running east–west and avenues running north–south. Students need not fully understand the meanings of the grid numbers at this stage. They will learn about degrees as units of measure for angles later in this unit.

3–4. Problems **3** and **4** focus on both map and real-world directions, encouraging students to draw connections between flat maps and the spherical Earth. While answering these questions, students may realize that there are problems representing Earth on a flat map. On the following pages, students investigate these problems in more detail and begin to realize the consequences of flat map depictions. The direction north plays an important role in these investigations.

A globe can help students examine and trace the planes' routes.

3. Students will revisit this question in problem **7** on the next page.

Your teacher has an orange. Imagine Earth as that orange, and the navel as the North Pole.

5. What problems might you have attaching the tiny map of the United States shown below to the orange?

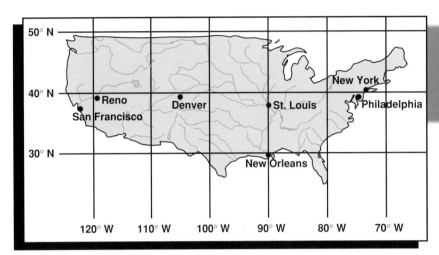

6. a. Suppose you cut off a piece of orange peel. What happens when someone tries to flatten the orange peel section on a table?

   b. What problems would a mapmaker have in making a map of Earth?

7. What is the connection between the separated piece of orange in the picture above, and the two planes flying north from Denver and Philadelphia?

Because Earth is a sphere and a map is flat, there are many different ways of drawing maps.

5. Answers will vary. Sample response:

It would be difficult to attach the map to the orange. Possible problems: the map might not adhere to the orange; the map might wrinkle since the orange is not flat like the map; the map might not curve to fit the shape of the orange.

6. a. Answers will vary. Sample response:

The skin may tear before it is flat on the table. Also, part of the orange peel might stick up.

b. Answers will vary. Sample response:

Flattening the round surface would distort the features of Earth.

7. Answers will vary. Sample response:

The planes are flying along the sides of the piece of peel. The planes will meet at the naval.

Materials orange (one per class); globe, optional (one per class); tennis ball, optional (one per class)

Overview Exploring the distortion of flat maps, students imagine attaching a flat map to an orange, as well as flattening a curved section of orange peel.

About the Mathematics The use of a flat model to represent a spherical surface distorts directions and distances, especially when larger surfaces of Earth are involved. On a local map, such as the map of Provo, Utah, from the previous section, a rectangular grid works; however, on the diagram of Utah on the next page, problems result from representing Utah as a rectangle.

Planning Students can work on problems **5–7** in pairs or in small groups. Discuss problems **6** and **7**.

Comments about the Problems

5. Students may draw the following conclusion: The smaller the map, the easier it is to attach it to a curved surface. Some students may now have insight into the kinds of distortion that are related to distances on maps.

6. Students should be able to understand the distortion of maps. If not, they may need to actually perform the orange peel activity. After students flatten their peels, they should place the peels orange-side-down on a table or desk. After a few seconds, the peels will start to curl. (You can also use a tennis ball and cut a section off for this demonstration.)

7. If students are having difficulty with problem **7,** a globe may help.

Writing Opportunity Have students write a paragraph explaining the difficulties of map-making.

Utah

When visitors travel to Utah from other states, they are welcomed at the state border by a sign that looks like a map of the state.

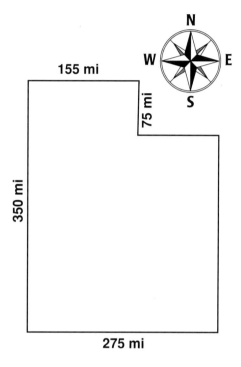

155 mi

75 mi

350 mi

275 mi

This diagram shows that Utah looks like a rectangle with a smaller rectangle cut from one corner.

8. Someone says the area of Utah is more than 100,000 square miles. Do you agree?

9. Align your pencil with north on the diagram. Does this represent *true* north for the people living in Utah? Explain.

Below are two different maps of Utah.

10. How are the shapes of the two maps different from the shape of Utah on the diagram above?

11. What differences do you notice between the two maps?

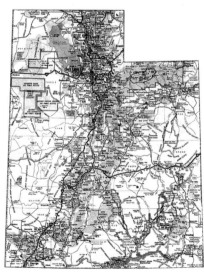

Map reduced from the Road Atlas © 1994 by Rand McNally.

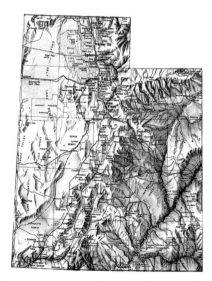

Map reduced from the New International Atlas © 1994 by Rand McNally.

8. No. If Utah were a complete rectangle, its area would be approximately 275 mi × 350 mi = 96,250 miles. However, Utah is smaller than that because there is a small rectangle cut out of the northeast corner.

9. No. North will be distorted (students should know this from the orange activity).

10. Answers will vary. Sample response:

The eastern and western borders of these two maps are not straight but slightly curved. When these borders are extended, they will eventually meet at the North Pole.

11. Answers will vary. Sample response:

The map on the right shows more contour to the land. Another difference is that the eastern and western borders in the map on the left are nearly parallel, whereas the borders seem to converge in the map on the right.

Overview Students investigate north on a diagram of Utah and compare two different maps of the state.

About the Mathematics The eastern and western borders of Utah have a north–south direction, so if they are extended they will meet like the airplanes in problem **3** and the edges of the orange peel in problem **7.** This fact helps to explain the difficulty in indicating true north on the different maps. If true north were indicated by arrows, then the arrows would be parallel when drawn on the diagram on the left below. They would not be parallel when drawn on the diagram on the right.

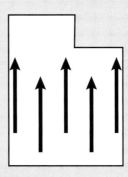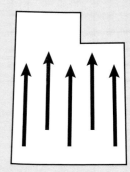

Planning Students can work on problems **8–11** in pairs or in small groups. Problem **8** is optional. Problem **11** can be used for assessment.

Comments about the Problems

8. Students' responses will give you an idea of their knowledge of the concept of area. Area is further discussed in the units *Reallotment* and *Made to Measure.*

11. Informal Assessment This problem assesses students' recognition of the different ways to present information in a map and the consequences of different presentation methods.

Writing Opportunity Invite students to investigate different maps and map-making (cartography) methods and to write about their investigations.

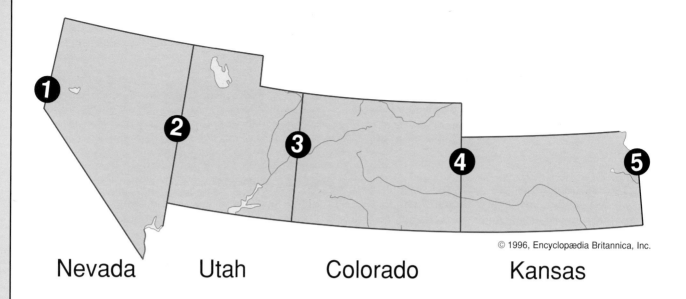

© 1996, Encyclopædia Britannica, Inc.

Nevada   Utah   Colorado   Kansas

This map shows four adjacent states: Nevada, Utah, Colorado, and Kansas.

12. Could you put a single arrow on the map that points to the north for all states?

13. Lay five pencils on the map to indicate the direction north for each of the locations 1, 2, 3, 4, and 5. What do you notice?

14. a. Using **Student Activity Sheet 5**, cut out the five small compasses and tape them onto the map at the positions 1, 2, 3, 4, and 5.

   b. Is it easy to find the directions north, south, east, and west? Why or why not?

   c. How can two people in different locations travel in the same direction and end up in the same place?

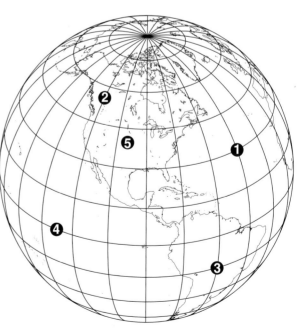

12. No, every state has numerous north arrows.

13. The pencils should be placed in line with the border lines. Students should notice that the lines will converge at the North Pole.

14. a. If the compasses are carefully placed, the north and south compass points will lie on lines of longitude, and the east and west compass points will lie on lines of latitude.

b. Answers will vary. One possible response is that it is not easy to find the directions because of the curvature of Earth.

c. Answers will vary. If two people start at the same line of latitude, they can both travel in the same direction, north or south, at the same speed, and end up in the same place (at the North or South Pole); or they can travel east or west at different speeds and end up in the same place.

If two people start at different lines of latitude, and both travel in the same direction, either north or south, at different speeds, they can meet at the North or South Pole.

Materials Student Activity Sheet 5 (one per student); transparency of Student Activity Sheet 5, optional (one per class); scissors (one pair per student); tape (one roll per pair or group of students); globe, optional (one per class)

Overview Students continue to investigate the direction north on a map. They also explore the four compass directions on a globe.

Planning Students can work on problems **12–14** in pairs or in small groups. You may want to have a class discussion after students finish problem **13**. Problem **14c** can be used for assessment.

Comments about the Problems

12–13. Have students reflect on these two problems in relation to problem **9** on the previous page.

14. a. If possible, this activity should also be performed using a globe. A flat map will not clearly show the convergence of the lines of longitude, nor the parallelism of the lines of latitude.

c. Informal Assessment This problem assesses students' ability to indicate a direction using cardinal directions. The problem provides an opportunity for students to reflect on the concepts they have learned so far in this section. They should now understand that the North and South Poles are fixed points, and that east and west are relative directions based on their relation to the poles.

Interdisciplinary Connection Make a geography connection by introducing the concepts of latitude and longitude.

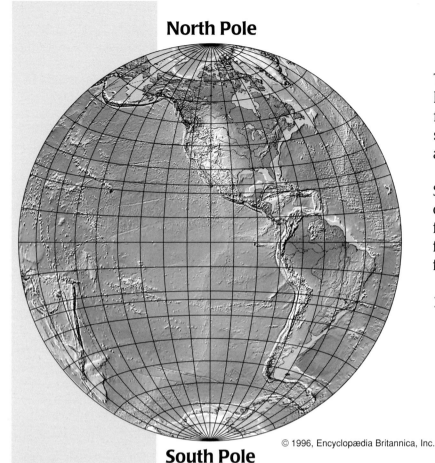

North Pole

© 1996, Encyclopædia Britannica, Inc.

South Pole

This map looks almost like the view of Earth from Apollo X. You can see the North Pole and the South Pole.

Suppose two planes take off at the same time, one from Denver and one from Philadelphia. They fly due south.

15. Could the planes ever meet? Explain. Compare your answer to the one you came up with on problem **3** (page 16).

Summary

It is impossible to make a perfectly accurate flat map of Earth.

Some maps have a grid and indicate the direction north with one arrow. Some maps show that Earth is a sphere—lines indicating north seem to go in slightly different directions.

For the rest of this unit, you will use simple grid maps.

Summary Questions

16. a. What is special about the equator?

   b. Describe the grid lines on a globe.

15. If they fly at the same speed, the planes will meet at the South Pole. In problem **3,** the two planes will meet at the North Pole rather than the South Pole.

16. a. Answers may vary. Sample response:

If you are on the equator, you are equidistant from the North and the South Poles.

b. Answers may vary. Sample response:

Lines of longitude run north–south. They meet at the North and South Poles. The distance between the lines of longitude is greatest at the equator. Lines of latitude lines run east–west. They never meet. The lines are a constant distance apart. They encircle Earth, running parallel to each other. The smallest circles are near the North and South Poles, and the largest circle is the equator.

Overview Students determine whether two planes headed due south will ever meet. They also read the Summary, discuss the uniqueness of the equator, and describe the grid lines on a globe.

Planning Have students read the Summary after they finish problem **15.** Problem **16** can be assigned as homework. After students complete Section C, you may assign appropriate activities from the Try This! section, located on pages 47–50 of the Student Book, as homework.

Comments about the Problems

15. Students are asked to point out the connection between this problem and problem **3** on Student Book page 16. Discuss with students the importance of the airplanes' departure points. You may also ask students to compare this map with the one on Student Book page 16.

16. Homework This problem may be assigned as homework. It offers students a writing opportunity.

Work Students Do

Students investigate headings using transparent compass cards. They determine courses for travel by plane and by boat. They also work informally with vectors, combining direction and distance to plot a hiking trip.

Goals

Students will:

- indicate a direction (heading) using degrees;*

- estimate and measure directions (headings) on a map or grid;

- use the scale on a map to estimate distances;

- use directions, headings, turns, and angles to solve simple problems;

- understand and use the relationships among directions, headings, and turns;*

- use directions, turns, and angles in combination with scales, distances, and the implicit use of vectors to solve more complex problems.

  * This goal is introduced in this section and is assessed in later sections of the unit.

Pacing

- approximately two to three 45-minute class sessions

Vocabulary

- compass card
- course
- degree
- distance
- heading

About the Mathematics

This section expands upon the eight wind directions by introducing a circular compass with 360°. *Headings,* using degrees instead of the eight wind directions, allow for a more precise description of direction. Headings are measured clockwise from north. Describing routes using headings and distances provides an informal introduction to vectors.

Materials

- Student Activity Sheets 6 and 7, pages 140 and 141 of the Teacher Guide (one of each per student)
- Transparency Masters 1 and 2, pages 146 and 147 of the Teacher Guide (one large and one medium compass card per student)
- rulers, pages 57 and 61 of the Teacher Guide (one per student)
- string, page 61 of the Teacher Guide, optional (one piece per student)
- paper strips, page 61 of the Teacher Guide, optional (one per student)
- U.S. maps, page 63 of the Teacher Guide (one per pair of students)

Planning Instruction

To introduce this section, you may want students to discuss the map of the San Francisco Bay area on page 21 of the Student Book. Ask students:

- *If a plane travels from the San Jose Airport to the Palo Alto Airport, what direction is it headed?* [northwest]

- *If a plane travels from the San Carlos Airport to the Oakland International Airport, what direction is it headed?* [Some students may say northeast; others may notice that the plane flies more north than east, or north–northeast.]

Students can work on problems 1–3 individually or in small groups. Problem 4 should be completed individually. Students can work in pairs on problems 5 and 13–16. Problems 6–12 can be done in pairs or in small groups.

There are no optional problems. Discuss each problem with the class. Elicit students' strategies and the reasoning used to solve each problem.

Homework

Problems 5 (page 54 of the Teacher Guide) and 10 (page 58 of the Teacher Guide) can be assigned as homework. After students complete Section D, you may assign appropriate activities from the Try This! section, located on pages 47–50 of the *Figuring All the Angles* Student Book. The Try This! activities reinforce the key mathematical concepts introduced in this section.

Planning Assessment

- Problem 10 can be used to informally assess students' ability to estimate and measure directions (headings) on a map grid and to use directions, headings, turns, and angles to solve simple problems.

- Problems 11 and 12 can be used to informally assess students' ability to use directions, headings, turns, and angles to solve simple problems and to use the scale on a map to estimate distances. They also assess their ability to use directions, turns, and angles in combination with scales, distances, and the implicit use of vectors to solve complex problems.

D. DIRECTIONS

San Francisco Bay Area

San Francisco is a city on the west coast of California. Here is a map of the San Francisco Bay area.

Use **Student Activity Sheet 6** to answer the problems below.

Golden Gate Bridge, San Francisco

1. A plane starts at San Carlos Airport and flies due west. Over which airport will the plane soon pass?

2. Another plane starts at San Carlos Airport and flies northwest. Over which airport will the plane fly first?

3. In which direction should a pilot fly to go from Hayward to Palo Alto?

Map from the Road Atlas © 1994 by Rand McNally.

1. Half Moon Bay

2. San Francisco International

3. almost directly south

Materials Student Activity Sheet 6 (one per student)

Overview Students use the eight wind directions to locate places on a map.

About the Mathematics These problems are linked to Section B, in which students used the eight wind directions on a compass. The next page introduces a compass card, which is used to describe directions more precisely.

Planning Students may work on problems **1–3** individually or in small groups. When students have finished problems **1–3,** encourage them to compare their answers.

Comments about the Problems

3. This problem can be used to challenge students to think about the precision of their answers. You may ask the class: *What can we do to make the compass direction heading more precise?* [Students may suggest further refining the compass by adding wind directions between the eight that exist, such as south–southeast. Others may suggest using degrees.]

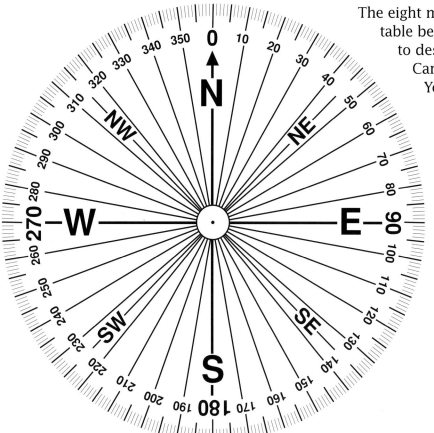

The eight main directions, shown in the table below, are not precise enough to describe a flight from San Carlos Airport to Hayward. Your teacher will give you a transparent **compass card.** The compass card can be used to describe the flight using **degrees.**

On the compass card you can see that:

- the direction north is the same as zero degrees (0°),

- east is 90°,

- south is 180°,

- and west is 270°.

4. Make a table like the one on the right in your notebook and fill it in.

5. Why would using the compass card with 360° be better for navigation than using just eight directions?

Compass Direction	Course (in degrees)
North	
Northeast	
East	
Southeast	
South	
Southwest	
West	
Northwest	

Solutions and Samples
of student work

Hints and Comments

4.

Compass Direction	Course (in degrees)
North	0°
Northeast	45°
East	90°
Southeast	135°
South	180°
Southwest	225°
West	270°
Northwest	315°

5. Answers will vary. Sample response:

Some places are not exactly north or northeast. You could describe their locations better using degrees.

Materials Transparency Masters 1 and 2 (one large and one medium compass card per student)

The masters for making the compass cards are included with the activity sheets at the end of this unit. You will need to make transparencies from the blackline master. Students should have their compass cards available for the rest of the unit.

Overview Students use compass cards to describe directions using degrees.

About the Mathematics Degrees are introduced with a circular compass having 360 degrees.

Planning Students can work individually on problem **4** and in pairs on problem **5.** You may want to assign problem **5** as homework.

Comments about the Problems

4. Since the eight compass directions are also indicated on the compass card, students should be able to read the degrees directly from the card. Make sure students can read the degrees on the compass card. Exactly between 0 and 10 degrees is the line indicating 5 degrees. Each small line represents 1 degree. You may want to point out, if necessary, that southwest is exactly between 220 and 230 degrees, at 225 degrees.

5. Homework This problem may be assigned as homework. If students are having difficulty, you might display a large map. Suggest a trip from one place to another and have students describe the travel directions using both wind directions and degrees. Ask: *Which of these two methods is more precise?* [using degrees]

Here is a way to set
a course from San Carlos
Airport to Hayward Airport.

First, put the center of
the compass card on San
Carlos Airport. Make sure
that N is pointing north.

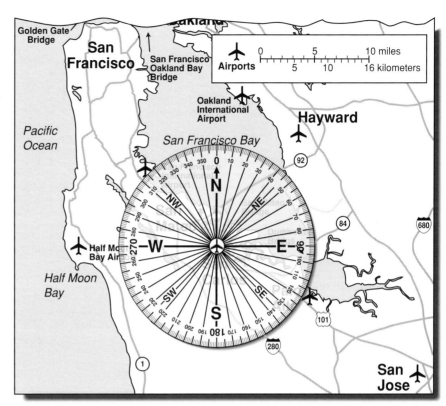

Map from the Road Atlas © 1994 by Rand McNally.

Second, place a ruler
from the center of the
compass card to
Hayward Airport.

Third, read the degree
mark at the edge of the
compass card. It is about
37°. This is called a
heading.

6. Use your compass
   card to determine
   the heading a pilot
   would fly to go
   from Oakland to
   Palo Alto.

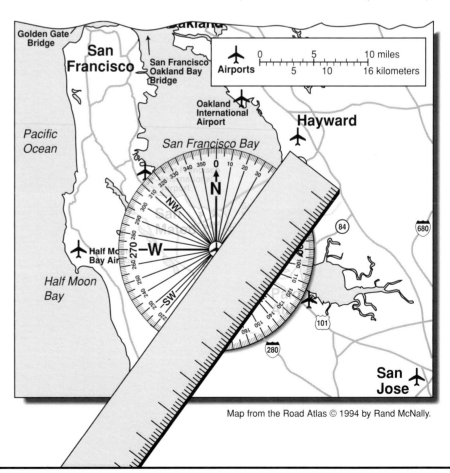

Map from the Road Atlas © 1994 by Rand McNally.

6. 160°

Materials Student Activity Sheet 6 (one per student); compass cards from Transparency Masters 1 and 2 (one large and one medium card per student); rulers (one per student)

Overview Students use their compass cards to determine headings.

About the Mathematics The direction north received so much attention in the previous sections because headings are always measured clockwise from north. Avoid calling a heading a *turn*; a turn is defined as a change in heading. For example, a left turn of 90° can result in a heading of 270°. Turns are introduced in Section F. Careful use of these terms can help avoid confusion.

Planning You may want to discuss this page with the whole class. Then students can work on problem **6** in pairs or in small groups.

Comments about the Problems

6. Students may need to work together to ensure correct use of the compass cards. It is important (and sometimes difficult) to keep the center point and north fixed. The center point of the compass card should be placed on the trip's starting point. Some students may find it helpful to draw lines heading north from the starting point on their maps.

See whether or not students understand how to find the correct heading in problem **6** before they continue with the problems on the next page. Students should understand that a heading is always measured clockwise from north. If students are having difficulty, ask them to return to the picture on this page and explain why the heading from San Carlos Airport to Hayward Airport is 37°. Then have them place their compass card on the map to find the heading from San Carlos Airport to another airport.

Two pilots planned flights from San Carlos to Oakland. Luiz, the first pilot, said, "It's exactly to the north." Ann, the other pilot, disagreed; she would fly the **course** heading of 10°.

7. Who is right?

They both arrived at Oakland Airport, but one pilot got there earlier than the other. They had to fly back after a short break. Luiz said, "I don't have to measure the heading back to San Carlos Airport. It is a heading of 190°."

8. Do you agree with Luiz? Explain your reasoning.

Captain Aziz and First Mate Mamphono are boating off the coast of California. They want to go to Sausalito Harbor.

9. Use **Student Activity Sheet 6** to draw a route to Sausalito. Harbor. Use only straight lines. Give the heading for each leg of the route.

You can sail to Sausalito using many different headings.

After arriving in Sausalito, Captain Aziz says, "I used only three different headings."

10. Is this possible? Explain.

Map labels: San Pablo Bay · 580 · 80 · Richmond San Rafael Bridge · Sausalito Harbor · Oakland · Golden Gate Bridge · San Francisco · San Francisco Oakland Bay Bridge · Oakland International Airport · Pacific Ocean · San Francisco Bay · San Francisco International Airport · San Mateo Bridge · San Mateo · Dumbart Bridge · Half Moon Bay Airport · San Carlos · Palo Alto · Half Moon Bay · Aziz & Mamphono · 1 · 280 · 9

7. Ann is right. The heading is about 10°.

8. Yes. Flying back means going in the opposite direction. According to the compass card, 190° is in the opposite direction of 10°.

9. Answers will vary. A possible route and corresponding headings are shown below.

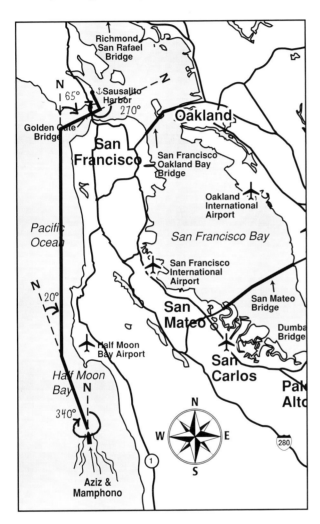

10. Yes. Aziz and Mamphono could head 340° for about 35 kilometers, change to a 55° heading and go under the Golden Gate Bridge, and then travel for another 2.5 kilometers and change their heading to 315°.

Materials Student Activity Sheet 6 (one per student); compass cards (one per student)

Overview Students use headings to solve simple navigation problems.

About the Mathematics Students may recall the concept of opposite directions from the unit *Patterns and Symbols,* in which they explored compass directions and their opposites. For example, east is the opposite of west.

Planning Students can work on problems **7–10** in pairs or in small groups. Discuss problem **8** with the whole class. Problem **10** can be assigned as homework and/or used for assessment.

Comments about the Problems

7. Students may have arrived at different answers depending on the accuracy of their measurements. For example, students may have incorrectly matched north on the compass card with north on the map, not positioned the compass card's center point directly over the middle of the picture of the airplane, or inaccurately read the degrees.

You may want to discuss these errors with your students and have them determine what happens when the compass card's center point is placed over one of the plane's wings. You can also agree upon a range of acceptable answers.

8. Some students may notice that going back means that the heading differs 180° from the original heading. If not, help students gain this insight by giving them another heading and then asking for the heading of the route in the opposite direction.

9. You may have students check each other's headings.

10. Informal Assessment This problem assesses students' ability to estimate and measure directions (headings) on a map or grid and to use directions, headings, turns, and angles to solve simple problems. If students only draw the route on the map, you might encourage them to describe the route by listing the heading of each leg of the trip. Some students may also describe the distance of each leg. This foreshadows the underlying concept of vectors. (See About the Mathematics on page 61.)

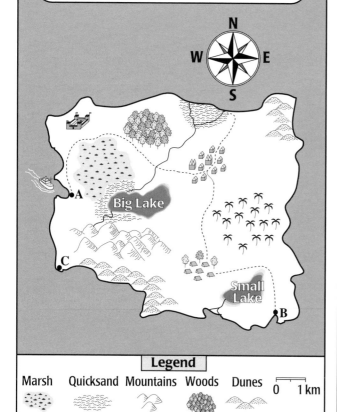

You and your friend are planning to hike on Sun Island. Use your compass card and a ruler to draw the paths described on the right. Use the map on **Student Activity Sheet 7.**

Legend

| Marsh | Quicksand | Mountains | Woods | Dunes |

0　　1 km

Your Trip

- Land at point A.

- Walk north 2 kilometers.

- Now walk due east to the closest edge of the woods.

- Now turn to a 160° heading.

- When you get to the river, swim across it.

- At the eastern bank of the river, head 70° for 1.7 kilometers and have a snack.

- Head 210° to get to the mountains.

- At the foot of the first mountain, you can rest. This is the end of your trip.

Your Friend's Trip

- Land at point A.

- Walk south 1 kilometer.

- From there, walk to the coast and travel along the coast to point C.

- Now plot a course 60° and walk nearly 2.5 kilometers, being careful to avoid the quicksand.

- Turn due east, and walk 4 kilometers to your destination.

11. **a.** How far did you travel?

　　b. How far did your friend travel?

12. When both of you have completed your trips, you decide to visit your friend. Plot the course and estimate the **distance** you need to travel to reach your friend.

11. a. about 10 kilometers

　　b. about 9.6 kilometers

12. about 3 kilometers

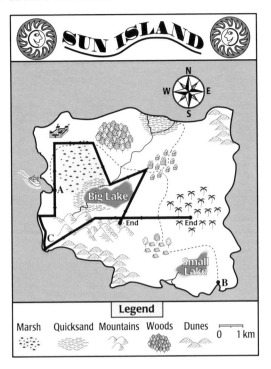

Materials Student Activity Sheet 7 (one per student); compass cards from Transparency Masters 1 and 2 (one large and one medium card per student); ruler (one per student); string, optional (one piece per student); paper strips, optional (one per student)

Overview Students plot the routes of two hiking trips on a map using both headings and distances.

About the Mathematics Each leg of the hike on Sun Island is a line segment that has a certain *direction* and represents a certain *distance*. This connection between direction and distance informally introduces students to vectors. In the next section, students continue to investigate vectors; however, the term *vector* need not be mentioned in this unit. Students will revisit the concept of vectors in the grade 8/9 unit *Going the Distance*.

These problems also involve changes in direction, foreshadowing the focus of Section F: changing directions and turns.

Planning Students can work individually or in pairs on problems **11** and **12**. You may want to use these problems for assessment.

Comments about the Problems

11–12. Informal Assessment These problems assess students' ability to use directions, headings, turns, and distances to solve simple problems and to use the scale on a map to estimate distances. They also assess their ability to use directions, turns, and angles in combination with scales, distances, and the implicit use of vectors to solve complex problems.

11. Students should organize their work carefully. First, have them determine the heading with their compass cards. Then have them draw a line in that direction. Finally, ask students to measure the correct distance. They may want to use string or paper strips for this activity.

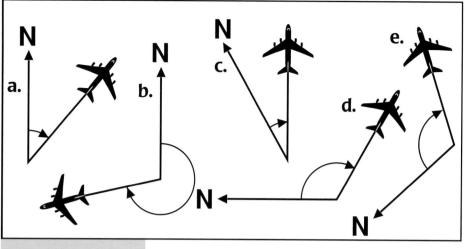

13. In which of the drawings on the left is the heading largest?

14. How many degrees apart are the following directions?

    a. north and east

    b. west and northeast

    c. south and southwest

    d. east and west

Summary

Together with distances, headings are used to plot courses. A *heading* is measured in degrees to the right from the direction north.

You can use your compass card to plot courses using degrees. North is 0°, east is 90°, south is 180°, and west is 270°.

Directions using degrees can be more accurate than the eight main directions. Degrees divide a circle into 360 parts, while the main directions divide a circle into only eight parts.

Summary Questions

15. What should be the heading of a plane flying from San Francisco to Salt Lake City, Utah?

16. Describe the headings pictured on the right.

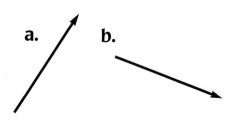

13. The heading is the largest for **b:** about 255°.

14. a. 90° or 270°

   b. 135° (90° + 45°) or 225° (180° + 45°)

   c. 45° or 315° (270° + 45°)

   d. 180°

15. The heading should be about 70°.

16. a. 33°

   b. 110°

Materials compass cards from Transparency Masters 1 and 2 (one large and one medium card per student), U.S. maps (one per pair of students)

Overview Students estimate and compare headings of airplanes in pictures for which north is not always up. Students describe changes in wind directions using degrees. They also read the Summary. Finally, students determine the heading of a plane flying from San Francisco to Salt Lake City, and describe two more headings using degrees.

About the Mathematics Estimating and measuring headings in pictures for which north is not "up," will improve students' understanding of angles and of measuring angles with a compass card, which are core parts of this unit.

Planning Students can work on problems **13–16** in pairs. After students complete Section D, you may assign appropriate activities from the Try This! section, located on pages 47–50 of the Student Book, as homework.

Comments about the Problems

13. Students can solve this problem by comparing and estimating. After they have solved the problem, you may want to ask them to measure these headings if they have not already done so. The pictures clearly show that headings are always measured to the right of the direction north. The larger the arrow's arc, the larger the heading.

14. You may ask your students to first attempt this problem without using their compass cards. If they are having difficulty, allow them to use their cards. Students will develop many interesting ways to find the change in heading. This problem previews the next section.

15. Students will need a suitable map to measure this heading. Have them align north on their compass cards with the line that forms the top of the border between California and Nevada.

SECTION E. NAVIGATION AND ORIENTATION

Work Students Do

Within the context of airports, students investigate polar coordinate systems. They locate airplanes on a radar screen (a polar grid) using distances and directions, and they participate in an air traffic control simulation. Students learn about the notations used to describe locations on a polar grid and on a rectangular grid. Finally, given two planes' locations (their headings and distances from the airport), students determine the distance between the planes.

Goals

Students will:

• identify a position using a polar grid;

• estimate and measure directions (headings) and distances on a map or grid;

• use directions, headings, turns, and angles to solve simple problems;

• compare rectangular and polar grid systems.

Pacing

• approximately two to three 45-minute class sessions

Vocabulary

• circular or polar grid

About the Mathematics

Students are introduced to the polar coordinate system used in navigation, such as air traffic control. A polar grid system is similar to a compass card with a central point and concentric circles that radiate out from that point. In a polar grid system, points are located using a heading (expressed in degrees) and a distance from a central point (expressed in miles or kilometers). For example, a flight controller might describe a plane's location and movement as 180°/20. This means that the plane is south of the control tower at a distance of 20 miles.

Materials

- Student Activity Sheets 8, 9, and 10, pages 142–144 of the Teacher Guide, (one of each per student)
- colored pencils, page 69 of the Teacher Guide, optional (one box per pair or group of students)
- radar screens of Styrofoam, page 71 of the Teacher Guide, optional (one per group of students)
- colored pawns, page 71 of the Teacher Guide, optional (one per student)
- compass cards from Transparency Master 2, page 73 of the Teacher Guide (one per student)
- LOST AT SEA, an activity on the software module *Maps and Navigation* from Sunburst Communications, Inc., page 75 of the Teacher Guide, optional (one per class)

Planning Instruction

Introduce this section with a discussion of air traffic controllers. You may want to ask:

What do air traffic controllers do? Why is their job important? [Students may say that air traffic controllers are like traffic cops; they direct traffic to ensure that airplane travel is smooth. Students may also say that air traffic controllers help to make airplane travel safe.]

How do air traffic controllers know where the airplanes are? [Students may know that air traffic controllers pay close attention to their radar screens, which show airplanes as small dots on circles.]

You may also want to share information about air traffic control in your local area. Students can work on the activity on page 29 of the Student Book in groups of five or more. They can work on problems 5, 6, 10, and 11 individually or in pairs and on problems 1–4 and 7–9 in pairs or in small groups.

There are no optional problems. Discuss each problem with the class. Elicit students' strategies.

Homework

Problem 5 (page 72 of the Teacher Guide), problem 7 (page 74 of the Teacher Guide), and problems 10 and 11 (page 76 of the Teacher Guide) can also be assigned as homework. After students complete this section, you may assign appropriate activities from the Try This! section, located on pages 47–50 of the *Figuring All the Angles* Student Book. The Try This! activities reinforce the key mathematical concepts introduced in this section.

Planning Assessment

- Problem 4 can be used to informally assess students' ability to identify a position using a polar grid.
- Problem 5 can be used to informally assess students' ability to compare rectangular and polar grid systems.
- Problems 6 and 7 can be used to informally assess students' ability to estimate and measure directions (headings) and distances on a map or grid and their ability to use directions, headings, turns, and angles to solve simple problems.

E. NAVIGATION AND ORIENTATION

TRAFFIC CONTROL

Some airports are very busy.

At Chicago's O'Hare International Airport, more than one plane lands and takes off every minute! During bad weather, this busy schedule leads to long lines of airplanes waiting for runways and many flight delays.

The *traffic control* takes place in a tower. Air traffic controllers try to organize traffic so that progress is smooth and travel is safe.

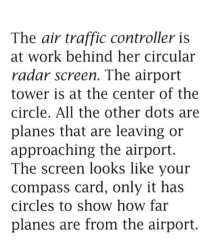

The *air traffic controller* is at work behind her circular *radar screen.* The airport tower is at the center of the circle. All the other dots are planes that are leaving or approaching the airport. The screen looks like your compass card, only it has circles to show how far planes are from the airport.

Overview Students read about air traffic controllers and their radar screens. This page establishes the context for the rest of the section. There are no problems to solve on this page.

About the Mathematics Students were informally introduced to the polar grid system used throughout this section when they solved problem **19** of Section B (designing a city with eight roads heading in the main wind directions). As shown on page 28 of the Student Book, a circular or *polar grid system* consists of a central point with several lines radiating out from this point that represent specific *headings* or directions. The concentric circles represent specific distances away from the central point.

Planning You may want to discuss the cities and roads students designed for problem **19** of Section B, as well as this section's main model: the circular radar screen.

You can also introduce students to common terms used by air traffic controllers. At busy airports, traffic on the ground is as heavy as it is in the air. The term *ground control* refers to the supervision of airplanes that are on the ground, taxiing or waiting to take off. *Local control* refers to the supervision of planes flying close to the airport (within 5 miles). *Approach control* refers to the supervision of more distant airplane traffic. In this section, students will solve problems involving approach control.

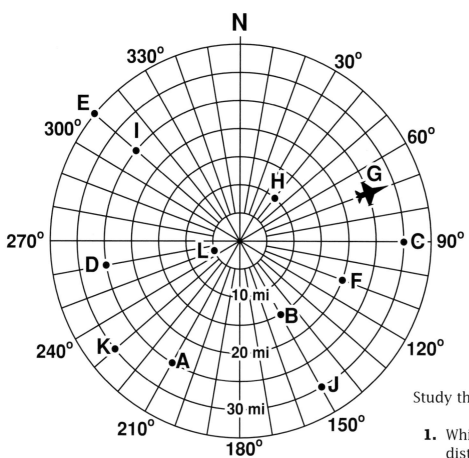

Here is a copy of such a radar screen. This type of grid is called a **circular** or **polar grid.**

On the grid on the left, the distance between each circle is 5 miles. Plane G, for instance, is 25 miles from the tower, at a heading of 70°. The notation 70°/25 miles is used to describe the location of plane G.

Study the radar screen.

1. Which planes are the same distance from the airport?

2. Which planes are in the same direction from the airport?

3. In your notebook, make a list of the airplanes and their locations as indicated on the radar screen. Your list should be similar to the list on the left.

Plane	Monitor
G	75° / 25 miles

A little later, the planes are in different locations.

Plane A   moved 10 miles closer to the airport.
Plane B   moved 5 miles farther away from the airport.
Plane C   is the same distance from the airport.
Plane L   landed.
Plane D   followed Plane L at 250° and is now 15 miles out.
Plane I   was told to go to a heading of 270° at 25 miles, but is only halfway there.
Plane G   is at 50°, 35 miles away.

4. Draw new positions for these planes on the radar screen on **Student Activity Sheet 8.**

1. Planes I, D, A, and G are all 25 miles from the airport. Planes K, J, and C are all 30 miles away.

2. Planes B and J are both at 150°. Planes E and I are both at 310°.

3.

Plane	Monitor
A	210° / 25 miles
B	150° / 15 miles
C	90° / 30 miles
D	260° / 25 miles
E	310° / 35 miles
F	110° / 20 miles
G	70° / 25 miles
H	40° / 10 miles
I	310° / 25 miles
J	150° / 30 miles
K	230° / 30 miles
L	250° / 5 miles

4.

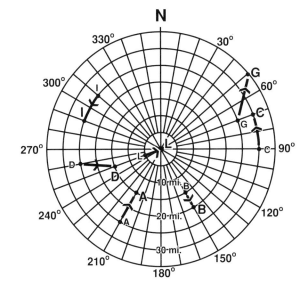

Materials Student Activity Sheet 8 (one per student); colored pencils, optional (one box per pair or group of students)

Overview Students locate airplanes on a radar screen using directions and distances.

About the Mathematics The concept of vectors is informally introduced when students use headings and distances to locate planes. Polar coordinates are also introduced.

Planning Start with a class discussion about radar screens. Ask: *What information does a radar screen give about airplanes? What information is not available from the screen?* [The radar screen shows the location of a plane as seen from a fixed point, the tower. The heading of the airplane can be determined only after the plane is spotted a second time.] Students can work in pairs or small groups on problems **1–4.** You may want to use problem **4** for assessment.

Comments about the Problems

1. Students can answer this question in different ways. They might mention the letters of the planes that are the same distance from the airport, or they may give a more general answer. For example, all planes on the same circle are at the same distance from the airport. If students do not suggest this more general description, you may ask:

   If planes are the same distance from the airport, where would you expect to find them on the screen? [The planes would be on the same circle.]

   What does it mean if planes are on the same circle? [The planes are the same distance from the airport.]

2. This question can also be answered by either naming the letters of the planes or by generalizing: all planes on the same radius are in the same direction from the airport.

4. **Informal Assessment** This problem assesses students' ability to identify a position using a polar grid. Colored pencils will help students keep track of the planes' new positions. You may also want to ask students to describe the new locations.

Activity

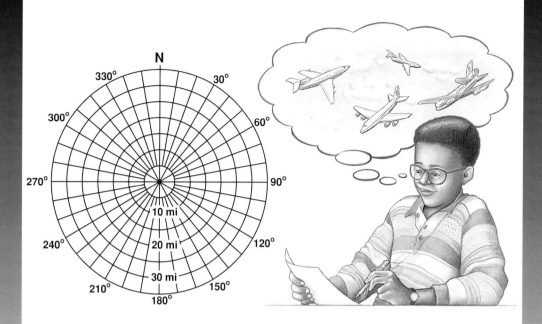

Work in groups of five or more students.

Use **Student Activity Sheet 9.**

One student in each group is the traffic controller. The others are pilots. Their planes are coming from different directions. To begin the game, pilots can select any place on the radar screen as their starting point.

- First, the pilots report their positions (heading/distance) to the traffic controller.

- The traffic controller then tells each pilot to proceed to a new position, one at a time. (A plane cannot cross more than one circle in a single move.) Repeat this step until all the pilots have safely landed their planes.

- There are two runways available: one east–west and one north–south.

- In order for a plane to land, no other plane can be within 10 miles of the airport. (No crashes, please!)

Materials Student Activity Sheet 9 (one per student); radar screens of Styrofoam, optional (one per group of students), colored pawns, optional (one per student)

Overview Students participate in an air traffic control simulation. In each group, one student is the traffic controller, and the others are pilots. The pilots report their positions to the air traffic controller, who then attempts to help each pilot land safely by giving appropriate directions.

Planning Divide students into groups of five or six. They might want to define rules for the activity. For example, students may decide to limit the distance a plane can fly along a circle in a single move. They may also decide to assign a wind direction to a main runway, thus limiting planes to landing with a headwind. Since the traffic controller is the game's organizer, it may be helpful for each group to appoint an assistant traffic controller. It is important for the traffic controller to arrange a sequential airplane lineup so that the planes do not arrive at the airport simultaneously. Have students change roles so that everyone has a chance to be both pilots and air traffic controllers.

5. Compare locating a position on a polar grid and locating a position on a **rectangular grid.**

There is a big fire in the Australian capital, Canberra. Chris is at location 1 on the map below. He phones his friend Tarin, who is at location 2. Chris says that he sees smoke in the direction 100°. Tarin says that from her vantage point, the smoke is at 40°.

6. Use **Student Activity Sheet 10** to locate the fire.

5. Answers will vary. Sample response:

The polar grid is good for situations in which traffic departs from and arrives at a central location. Both direction and distance are necessary to find a place on the polar grid.

The rectangular grid is better for representing streets that are parallel and perpendicular.

6. The location of the fire is within the circled region below.

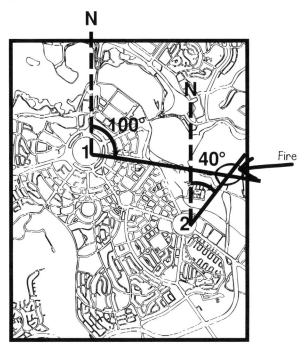

Materials Student Activity Sheet 10 (one per student), compass cards from Transparency Master 2 (one per student)

Overview Students compare the process of finding positions on a polar grid to that of finding them on a rectangular grid. They also draw headings on a map to locate a meeting spot.

About the Mathematics Different methods can be used to locate places on a flat map. In a rectangular coordinate system, a point is identified in terms of its distance from the horizontal and vertical axes. In a polar coordinate system, a point is identified in terms of its direction and distance from a fixed point. Problem **6** (and problem **7** on the next page) presents a third way to locate a point. Given the locations of two fixed points, a third location can be identified if the directions from each of the two fixed points to the third point are known.

Planning Students can work individually or in pairs on problems **5** and **6.** You may decide to use these problems for assessment. Problem **5** can be assigned as homework.

Comments about the Problems

5. Informal Assessment This problem assesses students' ability to compare rectangular and polar grid systems. Students may illustrate their descriptions with drawings.

6. Informal Assessment This problem assesses students' ability to estimate and measure directions (headings) and distances on a map or grid and their ability to use directions, headings, turns, and angles to solve simple problems.

To solve this problem, students must position the center of the compass card on location **1** and draw a line in the direction 100°. Then students should place the compass card on position **2** and draw a line in the direction 40°. The location of the fire is where the lines intersect.

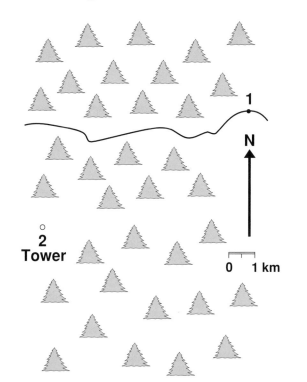

A plane has disappeared in a high forest. A ranger, driving his jeep on the road near location number 1, says he saw the plane disappear in the direction 200°.

The pilot's last report was that she saw the tower (location 2) in the direction west.

7. Where would you suggest that the rescuers look for the plane? Mark the place on the map on **Student Activity Sheet 10.**

Treasure at Sea

A treasure has been reported at the location 32W/33N on the map below. The harbor is at 36W/30N.

8. Determine what direct course (or heading) a ship will have to sail from the harbor to find the treasure.

A ship has a speed of 20 km/h.

9. Estimate how long it will take for the ship to reach the treasure.

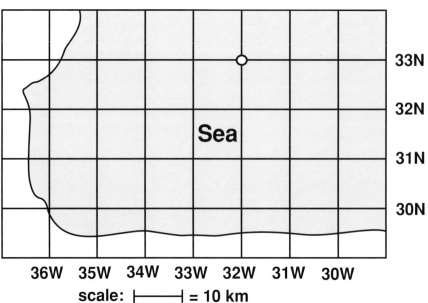

7. See diagram below.

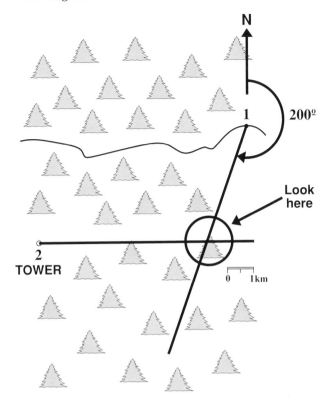

8. The ship must sail from the harbor at a heading of 53°.

9. The distance is about 50 kilometers, so it will take about 2.5 hours to get to the treasure.

Materials Student Activity Sheet 10 (one per student); LOST AT SEA software, optional

Overview Students use directions to locate a plane. They also plot a ship's course and estimate how long it will take for the ship to reach a sunken treasure.

About the Mathematics Students have learned that the notation 90°/50 miles indicates the location of a point (its direction/distance) in a polar grid. Problem **8** shows notation for a point's location in a rectangular coordinate system, such as 32W/33N. This notation will be revisited in the grade 6/7 unit *Operations.*

Planning Students can work in pairs or small groups on problems **7–9.** Problem **7** may be used for assessment or assigned as homework.

If you have access to the LOST AT SEA software, you may wish to give students time to explore triangulation, headings, angles, speed, and distance on the computer.

Comments about the Problems

7. Informal Assessment This problem assesses students' ability to estimate and measure directions and their ability to use directions, headings, turns, and angles to solve simple problems. This problem is similar to problem **6.** Students may reason as follows: If the pilot reported seeing the tower to the west, then she must have been somewhere to the east of location 2. So, one line can be drawn heading 90° from location 2, and a second line can be drawn heading 200° from location 1. The rescuers should look for the plane at the intersection of these two lines.

8. Do not confuse the notation 32W with degrees longitude. To solve this problem, students should plot 36W/30N and then use a compass card to plot the course to 32W/33N.

9. To estimate the time, students should first measure the distance from the harbor to the treasure and then use the map scale to find the distance in kilometers. Since the speed of the ship is given, they can now calculate: 50 ÷ 20 = 2.5 hours. This problem can also be solved using a double number line strategy:

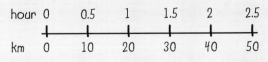

hour	0	0.5	1	1.5	2	2.5
km	0	10	20	30	40	50

Summary

Air traffic controllers use a polar grid system to identify the locations of planes.

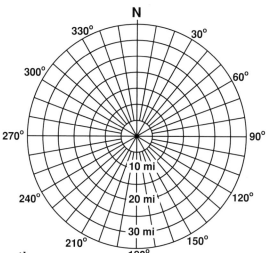

The system consists of two numbers:

- a heading expressed in degrees and

- a distance expressed in miles or kilometers.

A polar grid is different from a rectangular "city" grid. It has a central point from which the "roads" go in all directions, just like the degree lines on the compass card.

A flight controller describes a plane's location and movement as 90°/15 miles. This means the plane is east at a distance of 15 miles from the airport.

Summary Questions

One plane is at 90°/15 miles. Another plane is at 90°/35 miles.

10. What is the distance between the planes?

One plane is at 90°/15 miles. Another plane is at 270°/15 miles.

11. What is the distance between the planes?

10. The distance between the planes is 20 miles.

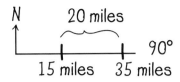

11. The distance between the planes is 30 miles.

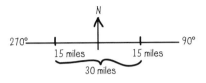

Overview Students read the Summary, which reviews the main topics of this and earlier sections. Given two airplanes' locations, students determine the distance between the planes.

Planning Problems **10** and **11** can be assigned as homework. Students can work on these problems individually or in pairs. After students complete Section E, you may assign appropriate activities from the Try This! section, located on pages 47–50 of the Student Book, as homework.

Comments about the Problems

10–11. Homework These problems may be assigned as homework. Encourage students to draw sketches. Students may also arrive at the answer by reasoning or by mentally plotting the planes on a polar grid.

10. The planes are headed in the same direction from the airport, but they are at different distances. Finding the difference between these distances yields the answer.

11. The planes are on the same circle. One plane is directly opposite the other (the difference in heading is 180°), so the two planes are located on a straight line. Adding the distances of both planes from the airport yields the answer.

Work Students Do

Students investigate turns, which are changes in a plane's direction. Assuming the roles of pilot, flight engineer, and traffic controller, students engage in an air traffic control simulation and model a plane's flight path. Students also explore the turns made by a dogsled. They follow and give directions for drawing several regular polygons. Finally, students investigate the angles that result from turns in both dynamic situations (navigation) and static situations (shapes).

Goals

Students will:

• estimate and measure turns and angles;

• use directions, headings, turns, and angles to solve simple problems;

• understand the relationships among turns, resulting angles, and the number of sides of a regular polygon;

• understand and use the relationships among directions, headings, turns, and angles;

• understand the relationship between the dynamic definition and the static interpretation of angles.

Pacing

• approximately four to five 45-minute class sessions

Vocabulary

• hexagon   • square

• pentagon   • triangle

• resulting angle   • turn

About the Mathematics

Vector drawings are used to estimate headings and describe turns. Vectors represent both magnitude (distance) and direction (heading). When two vectors are combined, they are drawn head to tail as shown on Student Book page 34. The turn needed to make a new heading is the angle formed by the extension of the old heading (shown as dotted line) and the new heading.

When two turns are both right or both left, the total turn is the sum of the individual turns. When combining a right turn and a left turn, the total turn is the difference between the two turns. The sums of the turns needed to travel around an equilateral triangle and a regular hexagon are also explored. The sum of the turns (exterior angles) needed to travel around any convex polygon is 360 degrees.

The angle formed by a turn is investigated. The sum of a turn and its resulting angle is 180 degrees. A regular polygon is a convex polygon in which all sides have the same length and all angles have the same measure. The size of the turn around a vertex of any regular polygon is $\frac{360}{n}$, where n is the number of sides or vertices. The interior angle (resulting angle) is the supplement of the turn, or $180 - \frac{360}{n}$.

Materials

- Student Activity Sheet 11, page 145 of the Teacher Guide (one per student)
- Transparency Master 3, page 148 of the Teacher Guide, optional (a few per class)
- compass cards from Transparency Master 2, pages 83, 85, 87, 91, 93, and 95 of the Teacher Guide (one per student)
- rulers, pages 83, 85, 87, 91, and 95 of the Teacher Guide (one per student)
- masking tape, page 85 of the Teacher Guide (one roll per class)
- meter stick, page 85 of the Teacher Guide, optional (one per class)
- grid paper, page 91 of the Teacher Guide, optional (one sheet per student)

Additional Resources

- a pilot or flight attendant as a guest speaker

Planning Instruction

You may want to introduce this section with the following activity: Ask a student to leave the classroom while the remaining students choose a location inside the room. Then invite the student who left back into the room and have the other students guide their classmate to the location with directions. For example, students might say *take three steps forward, turn 45° to the right,* and so forth, until the student reaches the destination.

Students can work on problems 1–3 as a whole class; 4–6, 9, and 24–25 individually; and problems 7, 8, 10–15, and 17–23 in small groups.

There are no optional problems. Discuss each problem with the class. Elicit students' strategies.

Homework

Problems 24 and 25 (page 94 of the Teacher Guide) can be assigned as homework. The Extension (page 91 of the Teacher Guide) can also be assigned as homework. After students complete this section, you may assign appropriate activities from the Try This! section, located on pages 47–50 of the *Figuring All the Angles* Student Book. The Try This! activities reinforce the key mathematical concepts introduced in this section.

Planning Assessment

- Problem 8 can be used to informally assess students' understanding of and ability to use the relationships among directions, headings, turns, and angles to solve simple problems.
- Problem 9 can be used to informally assess students' ability to estimate and measure turns.
- Problem 16 can be used to informally assess students' understanding of and their ability to use the relationships among turns, resulting angles, and the number of sides of a regular polygon.
- Problem 22 can be used to informally assess students' understanding of and their ability to use the relationships between turns and angles.
- Problem 23 can be used to informally assess students' understanding of and their ability to use the relationships among directions, headings, turns, and angles and to understand the relationship between the dynamic definition and the static interpretation of angles.

MORE TRAFFIC CONTROL.

A pilot uses many instruments to fly a plane safely.

On the left is a picture of a heading indicator. It shows the plane heading in a northern direction. The heading indicator looks different from your compass card.

On the heading indicator, the 3 means 30°, the 6 means 60°, and so forth.

1. According to this instrument, at what heading is the plane flying?

1. The plane is flying at a heading of 15°.

Overview Students examine an airplane's heading indicator to determine the plane's heading.

About the Mathematics Airplanes do not always maintain the same heading; sometimes they need to change directions, or make turns. This page connects the Sun Island activity from Section D (in which students drew hiking routes using different headings) and the air traffic control context from Section E.

Planning Students can do problem **1** as a class activity.

Comments about the Problems

1. To reinforce students' understanding of the airplane's heading indicator, ask: *What would the picture of the heading indicator look like if the plane were heading east?* [It would show the nose of the plane pointed toward the letter E.] The plane on the indicator does not move (it is painted on top of the glass), but the compass card that is behind the glass turns; it adjusts itself to indicate the proper heading.

A pilot is in radio contact with the control tower. The conversation between the air traffic controller and the pilot follows:

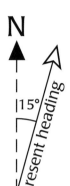

Traffic Control: Hello Flight 42. What is your present heading?

Pilot: Our heading is one-five degrees.

Traffic Control: Okay 42, make your heading three-five degrees.

Pilot: Our new heading will be three-five degrees.

Flight 42

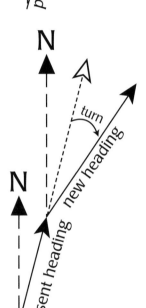

2. a. Using your compass card and a ruler, make a drawing of the present heading and the new heading. It should look something like the picture on the left.

The pilot of Flight 42 has to make a change in direction, or a **turn**.

b. Using degrees, describe the turn Flight 42 will make.

Here is another situation:

Traffic Control: Hello Flight 72. What is your present heading?

Pilot: Our heading is seven-five degrees.

Traffic Control: OK, 72, make your heading two-five degrees.

Pilot: Our new heading will be two-five degrees.

Flight 72

3. Describe the turn Flight 72 will make. Use a drawing to support your answer.

2. a. See the diagram below.

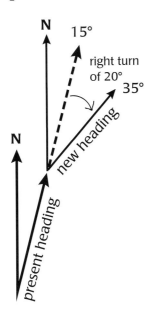

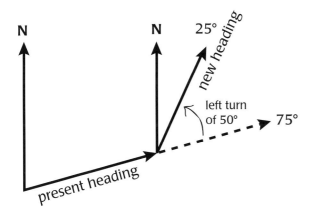

b. The plane will turn 20° to the right.

3. The plane will turn 50° to the left. See the diagram below.

Materials compass cards from Transparency Master 2 (one per student); rulers (one per student); Transparency Master 3, optional (a few per class)

Overview Students read the scripts of two conversations between an air traffic controller and a pilot. They then describe the airplanes' turns in degrees.

About the Mathematics So far in this unit, degrees have been used only in the context of directions or headings. The word *turn,* defined as a change in direction, is introduced on this page. A turn can be described using degrees. In fact, what we are dealing with here is an angle that can be measured. The word *turn* is used since it is tied to the dynamic airplane context. The term *angle* should not be used yet. It is used later in this section to describe the resulting angles in the tracks of a dogsled. Be very careful with terminology in this section so that students do not become confused.

Planning Students should work on problems **2** and **3** as a class, because their understanding of this page is crucial to the rest of the section. You may want to draw (or have a student draw) the path of Flight 42 on Transparency Master 3 to ensure that students grasp how the flight path can be sketched. Drawings should be as precise as possible.

Comments about the Problems

2. It may not be easy for students to make this kind of drawing. Students should position their compass cards with north facing up. The length of each line segment is not important; only the headings matter. At the point where the heading changes, the compass card should be placed north-up again. Have students write the headings in degrees. They should also indicate the turn with an arrow and the number of degrees.

3. Again, students practice sketching a turning plane's path. If students understood problem **2,** you may want them to solve problem **3** individually. You can then use this problem to check their understanding.

Activity

Clear an area in your classroom and designate an airport location. Choose three students to act out a scenario. One student acts as the pilot of an airplane, one acts as the flight engineer, and the other is the traffic controller.

The traffic controller gives the pilot a series of headings, one at a time. After each heading is given, the flight engineer tells the pilot how to turn. The pilot follows the directions and simulates the path of the flight by walking on the classroom floor. The rest of the class should use their compass cards to draw the path the plane follows to the airport.

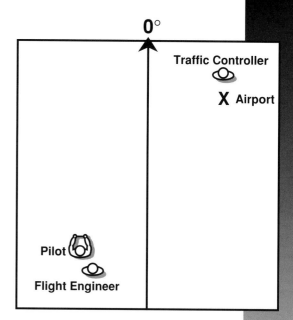

The pilot reports the plane's heading:

> **Pilot:** Flight 165 heading zero degrees.

The traffic controller requests a new heading:

> **Traffic Controller:** OK 165, make your new heading nine-zero degrees.

The flight engineer directs the plane to make a turn:

> **Flight Engineer:** Turn nine-zero degrees to the right.

The pilot walks three steps in the new direction. The flight engineer should use masking tape to keep a record of the plane's path on the floor.

The goal is to reach the airport, but not necessarily in the most direct way. After the exercise, discuss the flight path, paying attention to the headings and turns. Other members of the class may take over as pilot, flight engineer, or traffic controller and the entire process can begin again.

Materials compass cards from Transparency Master 2 (one per student); rulers (one per student); masking tape (one roll per class); meter stick (one per class)

Overview Students participate in an air traffic control simulation. One student acts as the pilot, a second student is the traffic controller, and a third is the flight engineer. After the traffic controller gives the pilot a new heading, the flight engineer tells the pilot how to turn, and the pilot must make the turn as precisely as possible. The goal of the simulation is to safely land the plane at a spot in the classroom designated as the airport.

About the Mathematics There are two ways to express a change in direction:

• indicate the new heading, from which the degrees of the turn can be determined using the difference between the old heading and the new heading or

• indicate how many degrees to turn left or right. In order to measure the angle of the turn, the N on the compass card should be lined up with the old heading.

Planning The problems on the previous page were from the onlooker's perspective. On this page, students actually engage in an air traffic control scenario, reinforcing their understanding of headings and turns. The whole class can participate in this activity, with students taking turns playing different roles. Allow students time to fully explore and understand headings and turns. Student spectators should be able to see and hear the simulation clearly. Ask these onlookers to describe what is going on. You may want to conduct the activity in a larger room in the school building.

The following conversation was taped in an airplane:

Traffic Control: Flight 33, what is your present heading?

Flight 33: Six-five degrees.

Traffic Control: Flight 33, make it one-zero degrees.

later...

Traffic Control: Flight 33, make it three-three-zero.

later...

Traffic Control: Flight 33, make it two-four-zero this time.

4. Draw the flight path.

5. The plane made three turns. Use your drawing to find the size of each turn.

6. The plane turned from a heading of 65° to a final heading of 240°. How large was the total turn?

Because a plane can make left and right turns, you cannot just add the degrees of two turns to find out the total turn. For example, if a captain first makes a right turn of 45° and then a left turn of 30°, the total turn is only 15° to the right. If he first makes a turn of 45° to the right and then another turn of 30° to the right, the result is a turn of 75° to the right.

A pilot was bothered that the air traffic controller asked her to make so many turns on approach.

"First they made me turn 30° to the right, then 20° to the left, then 15° to the left, then 10° to the right. Finally, I had to make a 5° turn to the left."

7. Explain why the pilot was so irritated.

A pilot made the following turns in this order: 20° right, 40° left, 45° left, 30° right, 10° left, 60° right. The original heading was 330°.

8. What was the heading after all the turns were made?

4. See the sample diagram below.

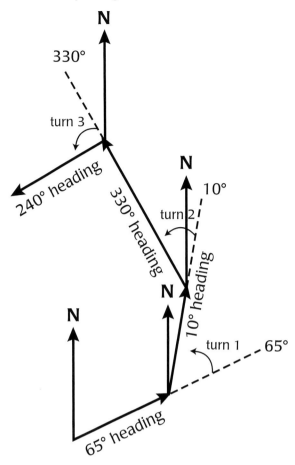

5. Turn 1: 55° to the left (or 295° to the right)

Turn 2: 40° to the left (or 320° to the right)

Turn 3: 90° to the left (or 270° to the right)

6. The total turn was 185° to the left.

7. The pilot was irritated because she made five turns but ended up going in the same heading.

8. The final heading was 345°. The turns resulted in a 15° right turn.

Materials compass cards (one per student); rulers (one per student)

Overview Students read the script of an air traffic controller's instructions to an airplane pilot. They then draw the plane's flight path, determine the sizes of its three turns, and compute its total turn. Given the original heading and a sequence of turns, students determine a plane's final heading.

Planning Students may work on problems **4–6** individually. Then invite them to discuss their answers in class. Students can work on problems **7** and **8** in small groups. You may decide to use problem **8** for assessment.

Comments about the Problems

4–5. Use these problems to check whether students can accurately draw a flight path involving headings and turns.

4. Students also drew the flight path of a plane in problems **2** and **3,** but now the situation involves a sequence of turns, which makes this problem more complex.

5. Students should use the compass card and align north with the old heading; the new heading indicates the number of degrees of the turn.

6. This problem summarizes the plane's turns, showing only its original and final headings. The total turn can be found by looking at the sketch drawn for problem **4** and adding the degrees for the three turns (55° to the left, 40° to the left, and 90° to the left; or 295° to the right, 320° to the right, and 270° to the right; or a mix of right and left turns). The total turn can also be found by comparing the original heading with the final heading. Using this strategy, some students may determine the total turn to be 175° to the right. This response would be correct if problem **6** were independent of problems **4** and **5.** Make sure students read and understand the text following problem **6.**

8. Informal Assessment This problem assesses students' understanding of and their ability to use the relationships among directions, headings, turns, and angles to solve simple problems.

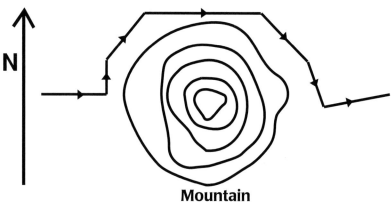

Mountain

In Antarctica, a dogsled is going from west to east. The sled has to make a detour around a mountain.

9. The sled makes six turns. The turns are sketched in the diagrams below. Estimate the size of each turn.

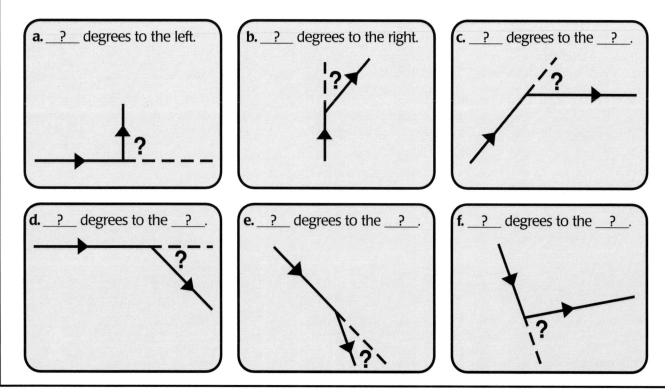

a. __?__ degrees to the left.

b. __?__ degrees to the right.

c. __?__ degrees to the __?__.

d. __?__ degrees to the __?__.

e. __?__ degrees to the __?__.

f. __?__ degrees to the __?__.

9. a. 90° to the left

   b. about 40° to the right

   c. about 45° to the right

   d. about 50° to the right

   e. about 25° to the right

   f. about 80° to the left

Overview Students study the tracks of a dogsled and estimate the sizes of its turns.

About the Mathematics The context of the dogsled will be revisited later in this section to introduce the concept of a *resulting angle.* In this context, a *resulting angle* is the angle formed by the parts of the track. On this page, however, the focus is on the turn. To avoid confusion, do not mention the term *angle.*

Planning Students can work on problem **9** individually. You may also use this problem for assessment.

Comments about the Problems

9. Informal Assessment This problem assesses students' ability to estimate turns. Encourage students to discuss their strategies. Have students estimate the sizes of the turns instead of measuring with a compass card. They should know what a 90° turn looks like from their work with compass cards. If students do not know how to estimate the size of a turn, ask them if the turn is larger or smaller than 90°. There is no need to introduce the terms *right, obtuse,* and *acute* angles. These terms will be discussed in Section G.

Compass Card Drawings

A ship has to go up the river in thick fog. The skipper chooses to sail only in two directions: 0° and 45°.

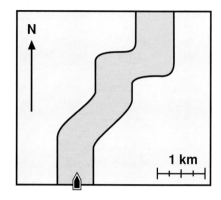

1 km

10. Construct a possible route on the map on **Student Activity Sheet 11.**

11. Give commands to the skipper (for instance, 1.5 kilometers ahead, 45° to the right) so that he can sail your route.

12. a. Follow the instructions on the right to make a drawing on a separate sheet of paper. Use your compass card and a ruler.

What figure have you just drawn?

b. How many degrees did you turn altogether?

- Go 2 centimeters north.
- Turn 90° to the right.
- Go 2 centimeters ahead.
- Turn 90° to the right.
- Go 2 centimeters ahead.
- Turn 90° to the right.
- Go 2 centimeters ahead.
- Turn 90° to the right.

13. Give instructions similar to those in problem **12** that will result in the two drawings on the right.

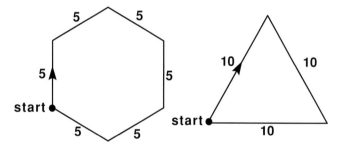

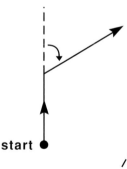

On the left is the first turn made in the drawing of the ***hexagon.***

14. How many degrees do you turn altogether when making the hexagon?

15. How many degrees do you turn altogether when making the ***triangle?***

16. What do you notice about your answers to problems **12b, 14,** and **15?** How can you explain this result?

10. Routes may vary. Sample route:

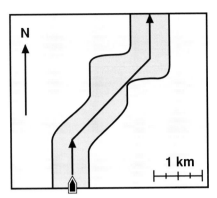

11. Answers will vary. Commands should be like these:
1 kilometer ahead, 45° right or starboard,
2.5 kilometers ahead, 45° left or port.

12. a. The directions make a square with sides
2 centimeters long.

   b. 360° (four turns to the right, each 90°)

13. Hexagon:
Go 5 centimeters ahead.
Turn 60° to the right.
Go 5 centimeters ahead.
Turn 60° to the right.
Go 5 centimeters ahead.
Turn 60° to the right.
Go 5 centimeters ahead.
Turn 60° to the right.
Go 5 centimeters ahead.
Turn 60° to the right.
Go 5 centimeters ahead.
Turn 60° to the right.

Triangle:
Go 10 centimeters ahead.
Turn 120° to the right.
Go 10 centimeters ahead.
Turn 120° to the right.
Go 10 centimeters ahead.
Turn 120° to the right.

14. 360° (6 × 60° = 360°)

15. 360° (3 × 120° = 360°)

16. They are the same. Students may say that in
traveling completely around each shape, you must
make a 360° turn.

Materials Student Activity Sheet 11 (one per
student); compass cards from Transparency
Master 2 (one per student); rulers (one per
student); grid paper for drawing polygons,
optional (one sheet per student)

Overview Students give and follow directions
involving turns to make geometric figures.

About the Mathematics Students move from
dynamic navigation problems to static problems
involving shapes. However, these problems still
focus on the turns, not on the resulting angles (or
interior angles) of the shapes.

Planning Students can work on problems **10–15**
in small groups. You may want to use problem **16**
for assessment.

Comments about the Problems

10–11. These problems connect the navigation
situations to those involving shapes.
Students should give directions using
turns, and they should indicate distances
based on the map scale.

   12. Encourage students to make precise
drawings.

   13. In order to find the size of a turn, some
students may need to extend the line
indicating the original heading. They may
check their answers by using their own
directions to draw the figures.

14–15. The more accurately students measure
the turn, the closer their answers will be
to 360°. Encourage students to explain
why the total of the turns is 360°.

   16. Informal Assessment This problem
assesses students' understanding of and
their ability to use the relationships among
turns, resulting angles, and the number of
sides of a regular polygon. Encourage
students to sketch a drawing in which all
turns are indicated with arrows. This will
help them see that they will get a full
circle if all these arrows are put together.

Extension Ask students to design their own
problems by writing directions for drawing
regular or irregular shapes.

Turns and Angles

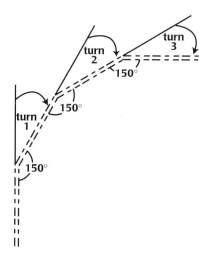

A sled got lost in the darkness of a polar night. "Mayday" emergency calls were received all night, but darkness prohibited a search. The next morning, planes searched the area, and the pilots saw tracks made by the sled.

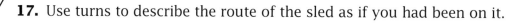

17. Use turns to describe the route of the sled as if you had been on it.

18. If the sled continued in the same way, it might have returned to the starting point. How many turns would the sled have had to make to return to the starting point?

The pilot describes the tracks as follows:

"It looks like the sled made three equal turns to the right. The four parts of the track seem equally long and the **resulting angle** between each part measures about 150°."

19. a. What do you think the pilot means by the *resulting angle*?

    b. How is the pilot's description different from the one you made in problem **17?**

20. a. If you were on a sled and made a turn of 40°, what would be the resulting angle between parts of the sled tracks?

    b. If parts of the sled tracks form a resulting angle of 130°, what is the size of the turn?

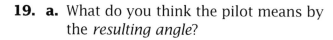

17. Students' answers will vary. However, their responses should focus on the four equal segments (expressed in a number of steps, meters, kilometers, or some other unit) and three equal turns of 30°.

18. The sled would have to make 12 equal turns of 30°.

19. a. the angle between the track segments

    b. The pilot used the angle visible in the tracks instead of turns.

20. a. 140°

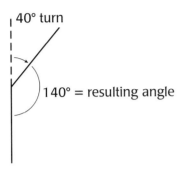

    b. 50°

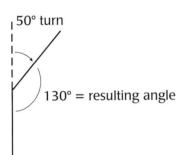

Materials compass cards from Transparency Master 2 (one per student)

Overview Students study a dogsled's tracks through the snow and describe the sled's route. They also examine the resulting angles between the parts of the track. Finally, given a resulting angle, students find the turn, and vice versa.

About the Mathematics The term *resulting angle* is formally introduced on this page. The resulting angle is the angle formed by the segments of the sled tracks. A resulting angle can also be called an interior angle, and a turn can be called an exterior angle. However, to avoid confusion, do not refer to a turn as an angle.

Also, in constructing drawings it is important to distinguish between the symbols that are used: an arrow is used to indicate a turn and an arc is used to indicate a resulting angle.

Planning Students may work on problems **17–20** in small groups.

Comments about the Problems

17. You may refer to the activity on page 35, where students modeled turns of an airplane. Some students may need to model the movements of the sled. Encourage them to draw the sled's turns.

18. Some students may notice a pattern related to the number of turns around the figure, the number of sides the figure has, and the total turn of 360°. (See problem **16** on the previous page.)

19. Use this problem to check whether students understand the difference between a turn and a resulting angle. They can use a drawing to show the difference.

20. Students should begin to recognize that a turn and a resulting angle together form a straight line (or a 180° angle). Encourage students to make drawings.

21. Why is it somewhat strange to talk about an angle of 180° between two parts of a track? Draw an angle of 180°.

22. What size turn do you make to get a resulting angle that is equal to the turn?

23. Look at the figures below. Think of their outlines as sled tracks.

a. Describe the turns that you have to make in each case.

b. How large are the resulting angles? (The resulting angles are shown with question marks.)

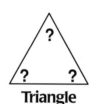
Triangle

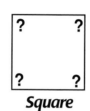
Square

Pentagon

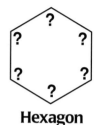
Hexagon

Summary

If you change from one direction to another, you make a turn. Turns can be made to the right or to the left.

For example, if you change from a heading of 30° to a heading of 45°, you make a turn of 15° to the right.

When you make a turn, your path forms a resulting angle.

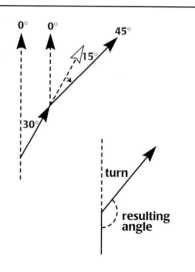

Summary Questions

24. In your own words, describe the relationship between a *turn* and the *resulting angle.*

25. a. If a turn is 100°, what is the resulting angle?

b. If a turn is 10°, what is the resulting angle?

21. An angle of 180° is a straight line.

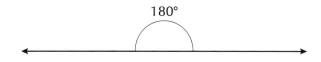

180°

22. A turn of 90° creates a resulting angle of 90°.

23. a. triangle   120° turns (360° ÷ 3)

       square   90° turns (360° ÷ 4)

       pentagon   72° turns (360° ÷ 5)

       hexagon   60° turns (360° ÷ 6)

   b. triangle   60° (180° − 120°)

       square   90° (180° − 90°)

       pentagon   108° (180° − 72°)

       hexagon   120° (180° − 60°)

24. Explanations will vary. Sample response:

A turn and the resulting angle add up to 180°.

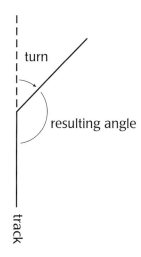

25. a. 80° (180° − 100°)

   b. 170° (180° − 10°)

Materials compass cards (one per student); rulers (one per student)

Overview Students solve problems that involve the relationship between turns and resulting angles, and they describe this relationship in their own words.

Planning Students can work on problems **21–23** in small groups. Problems **22** and **23** can be used for assessment. After reading the Summary, students can complete problems **24** and **25** for homework. After students complete Section F, you may assign appropriate activities from the Try This! section, located on pages 47–50 of the Student Book, as homework.

Comments about the Problems

21. Encourage students to think about the problem before they draw using their compass cards.

22. Informal Assessment This problem assesses students' understanding of and their ability to use the relationships between turns and angles. Students may use their compass cards if they have difficulty solving this problem.

23. Informal Assessment This problem assesses students' understanding of and their ability to use the relationships among directions, headings, turns, and angles and their understanding of the relationship between the dynamic definition and the static interpretation of angles.

You may refer to problems **12–16** in this section, where students determined the degrees of the turns needed to draw some of these shapes. Students should now recognize that a turn and its resulting angle add up to 180°.

24. Homework This problem may be assigned as homework. Students may also illustrate their descriptions.

25. Homework This problem may be assigned as homework. Again, students can use the fact that the sum of a turn and its resulting angle is 180°.

Work Students Do

Students explore the angles in mosaic tiles and determine whether the angles are *acute, right*, or *obtuse*. They study a map of a city whose streets extend from a central point like wheel spokes. Students determine the size of the angle formed by two adjacent roads. Finally, they explain how to use a compass card to measure a static angle.

Goals

Students will:

• use directions, headings, turns, and angles to solve simple problems;

• estimate and measure angles.

Pacing

• approximately two 45-minute class sessions

Vocabulary

• acute angle

• edge

• obtuse angle

• right angle

About the Mathematics

This section focuses on static angles in shapes and patterns without reference to movement (such as a turn or heading). Static angles are found in art and nature. The terms *acute, right,* and *obtuse* describe the sizes of angles. An acute angle has a measure less than 90°; a right angle has a measure of 90°; and an obtuse angle has a measure greater than 90° but less than 180°.

Materials

- mosaic tiles of the shapes featured in this section, pages 99 and 103 of the Teacher Guide, optional (one set per group of students)

- compass cards, optional, pages 99 and 105 of the Teacher Guide (one per student)

- tracing paper, page 103 of the Teacher Guide, optional (one sheet per student)

- scissors, pages 103 and 109 of the Teacher Guide, optional (one pair per student)

- copies of Transparency Masters 4, 5, and 6, pages 149–151 of the Teacher Guide (for an optional activity on page 109 of the Teacher Guide), optional (one of each per student)

Planning Instruction

Introduce this section with a discussion of angles in the classroom. Ask students:

Do you see any angles in the classroom? Can you describe them? [Students may notice the edges of a desk, the sides of a chalkboard, the hands of a clock, and so on.]

Students can work on problems 2, 3, 10, and 11 individually or in small groups and on problems 1 and 4–9 in small groups.

Homework

Problem 10 (page 106 of the Teacher Guide) and problem 11 (page 108 of the Teacher Guide) can be assigned as homework. The Extension activities (pages 105 and 109 of the Teacher Guide) can also be assigned as homework. You may also challenge students to find objects or pictures of objects with interesting angles for homework. After students complete this section, you may assign appropriate activities from the Try This! section, located on pages 47–50 of the *Figuring All the Angles* Student Book. The Try This! activities reinforce the key mathematical concepts introduced in this section.

Planning Assessment

- Problem 4 can be used to informally assess students' ability to use directions, headings, turns, and angles to solve simple problems.

- Problem 10 can be used to informally assess students' ability to estimate and measure angles.

G. ANGLES AND SHAPES

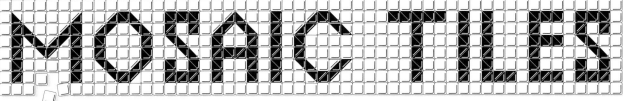

In this section, you will look at shapes and angles in those shapes. All of the shapes will be bounded by straight lines called **edges.** The edges form angles. These angles tell us a lot about the shapes, and the shapes tell us a lot about the angles.

An angle of 90° is called a **right angle.**

1. Where do you see right angles in the classroom? Can you form a right angle with your arm or leg?

The mosaic tile below has one right angle.

2. Which angle in the above tile measures 90°?

3. What can you say about the measures of the other angles in the above tile? (Hint: Look at the shapes below.)

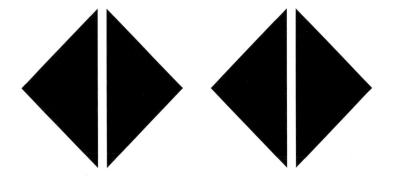

1. Answers will vary. Possible answers include: corners of books, tables, walls, floor tiles, and so forth.

   Yes, you can form a right angle with your arm or leg.

2. the top one

3. The other angles are equal in size. When placed next to each other, the right and left angles equal the top one, which measures 90°. Therefore, the right and left angles each equal 45°.

Materials mosaic tiles of the shapes featured in this section, optional (one set per group of students), compass cards, optional (one per student)

Overview Students identify right angles in the classroom and in mosaic tiles.

About the Mathematics This page focuses on the static interpretation of angles. There are no turns involved, just the resulting (interior) angles in shapes.

Planning Use problem **1** to introduce this section. Students can work on problems **2** and **3** in small groups or individually.

Comments about the Problems

1. Some students may have difficulty finding right angles in the classroom. When students point out right angles, ask them to explain why they think the object's edges form a right angle.

2. If students are having difficulty deciding which angle measures 90° have them use a compass card.

3. Students may also notice that the two smaller angles add up to 90° or that two of these triangular tiles form a square. Investigating real tiles may lead students to discover more ideas about the angles in a triangle.

An angle less than 90° is called an *acute angle.*

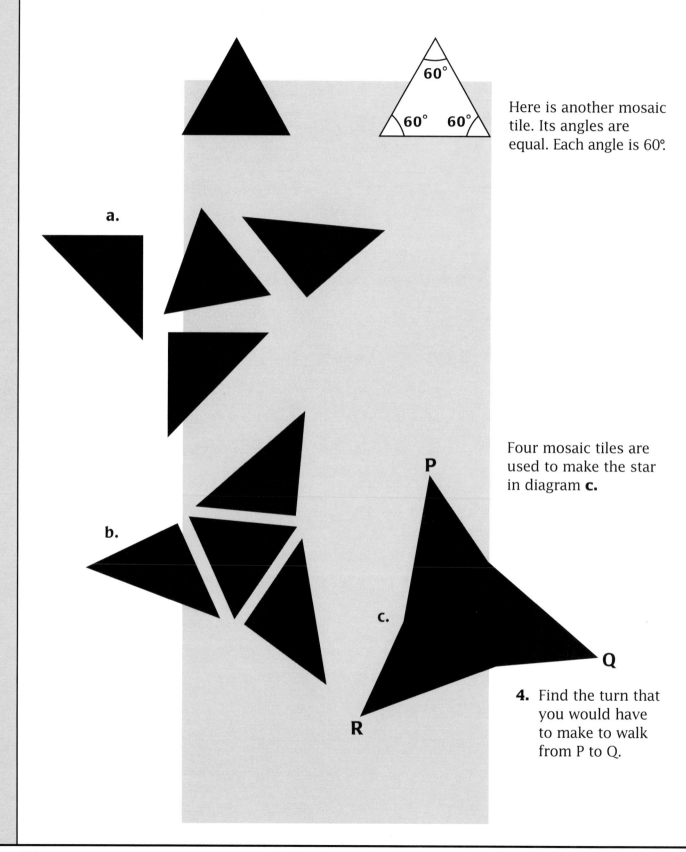

Here is another mosaic tile. Its angles are equal. Each angle is 60°.

a.

b.

c.

Four mosaic tiles are used to make the star in diagram **c.**

4. Find the turn that you would have to make to walk from P to Q.

4. Answers will vary, depending on the perspective of the student. One possible answer: The turn is 15° to the left if the student walks along the edge of the triangle toward the center triangle.

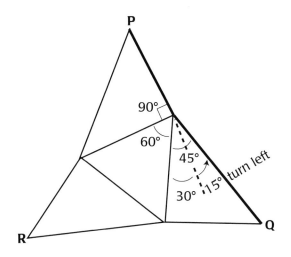

Overview Students continue their investigation of the angles in mosaic tiles. They use their knowledge of the relationship between turns and resulting angles to find an unknown turn.

About the Mathematics Students often have difficulty extending the lines of a figure and walking around figures. However, these activities are necessary for developing students' understanding of perimeter, complementary and supplementary angles, and construction strategies for complex shapes.

Planning Students can work on problem **4** in small groups. You may also use this problem for assessment. Have students discuss their answers in class.

Comments about the Problems

4. Informal Assessment This problem assess students' ability to use directions, headings, turns, and angles to solve simple problems.

This problem is similar to those in Section F in which students investigated directions for drawing polygons. Encourage students to make a drawing and label the known angles in the drawing. They should determine the size of the turn by reasoning, not measuring. For example, a student may think: I know that the three angles that meet at the turning point measure 90°, 60°, and 45°, for a total of 195°. If I extend the line that starts at **P,** I know that the measure of the angle formed is 180°. So the turn is 195° −180°, or 15°.

So far you have seen names for angles that are less than 90° (acute angles) and exactly 90° (right angles). Angles that are greater than 90° and less than 180° also have a special name. They are called **obtuse angles.**

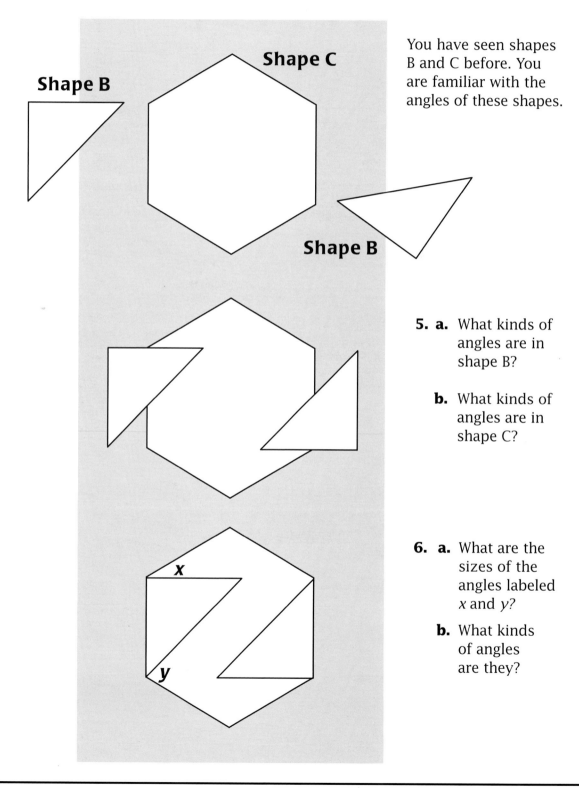

You have seen shapes B and C before. You are familiar with the angles of these shapes.

5. a. What kinds of angles are in shape B?

   b. What kinds of angles are in shape C?

6. a. What are the sizes of the angles labeled *x* and *y*?

   b. What kinds of angles are they?

5. a. one right angle and two acute angles

b. six obtuse angles

6. a. The angle labeled *x* is 30°, and the angle labeled *y* is 75°. See the diagrams below.

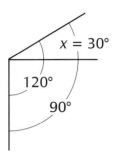

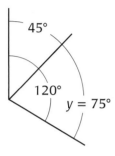

b. Both angles are acute.

Materials mosaic tiles of the shapes featured in this section, optional (one set per group of students); tracing paper optional (one sheet per student); scissors, optional (one pair per student)

Overview Students determine whether angles are *acute*, *right*, or *obtuse*. Then they use reasoning to determine the size of an angle.

Planning Students can work on problems **5** and **6** in small groups. You may want students to share their answers in class. Encourage them to find the answers to these problems by reasoning. Discourage students from using their compass cards to measure the angles.

Comments about the Problems

5–6. Invite students to explore the shapes and angles by tracing and cutting them out.

5. a. There is one right angle. The two other angles are equal. Since the three angles add up to 180°, each of the smaller angles is 45°.

b. You may refer to Section F, in which students explored turns and resulting angles for a hexagon. The only new concept here is the term *obtuse*.

6. To find the angle labeled *x*, subtract 90° from 120°. For the angle labeled *y*, subtract 45° from 120°.

Palmanova

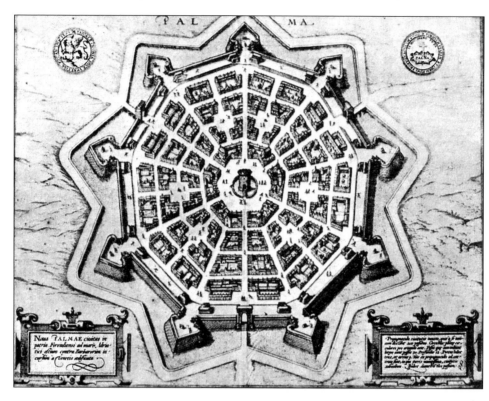

Above is the copy of the plan for Palmanova, a 16th-century city in Italy. You can see six major roads on the map of Palmanova that extend all the way into the central plaza.

7. What are the headings, in degrees, of these major roads?

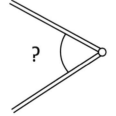

8. How large is the angle between two adjacent major roads?

There are two smaller roads between each major road.

9. How large would the angle between the two smaller roads be?

7. The headings are 0°, 60°, 120°, 180°, 240°, and 300°.

8. 60°. Students may measure or may reason as follows: The sum of the angles is 360°, and there are six equal angles, so each must measure 60°.

9. 20°. There are two smaller roads between each major road, forming three equal angles, so each must measure 20°.

Materials compass cards, optional (one per student); artwork or other illustrations of geometric shapes and angles, optional (several per class)

Overview Students study a map of Palmanova, an ancient Italian city designed like a polar grid system. Students determine the headings of the major roads and find the angles between adjacent roads.

About the Mathematics It is often possible to use reasoning to determine the size of an angle. Sometimes, however, an angle must be measured. When using a compass card to measure an angle, students must align north with one of the lines forming the angle. The other line then indicates the size of the angle.

Planning Students can work on problems **7–9** in small groups. If they are having difficulty with the problems, you may want them to discuss their answers.

Comments about the Problems

7. Some students may use their compass cards to measure the headings; others may use reasoning to determine the headings.

8. Again, the answer can be found by reasoning.

9. The angle between two small roads is formed by extending the roads into the center of the city.

Extension You may want to find and display artwork or other illustrations that contain geometric shapes and angles. Ask students to use reasoning or their compass cards to determine the sizes of angles in each picture.

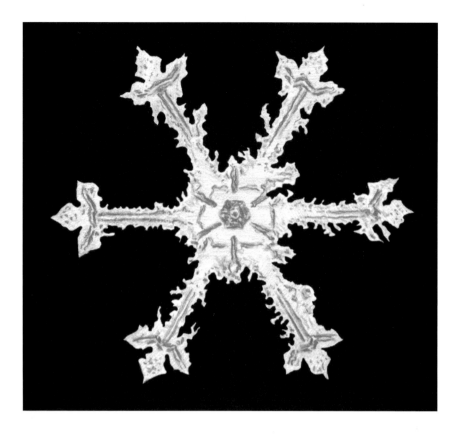

The city plan of
Palmanova has the
same structure as
a snowflake.

The snowflake forms a
six-pointed star.

Starfish often form
a five-pointed star.

10. About how large
is the angle
between two
adjacent legs of
the starfish?

10. About 72°. There are 360° in a circle. There are five equal angles within the imaginary circle around the starfish, so each angle is about 72°.

Overview Students study angles formed in a snowflake and a starfish.

Planning You may want to use problem **10** for homework or assessment. Students can work individually on this problem.

Comments about the Problems

10. Informal Assessment This problem assesses students' ability to estimate and measure angles. It can also be assigned for homework. Students should use reasoning to determine the size of the angle.

Interdisciplinary Connection Nature displays a rich variety of shapes and angles. You may want to extend this unit's concepts with a science class investigation of the geometry that can be found in nature.

Summary

Angles play an important role in many geometric figures. Sometimes you can find the size of an angle by reasoning, sometimes by estimating, and sometimes by measuring.

Angles come in different sizes:

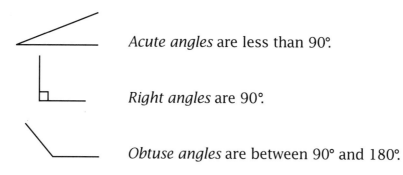

Acute angles are less than 90°.

Right angles are 90°.

Obtuse angles are between 90° and 180°.

Summary Question

A student placed a compass card on an angle and said, "It looks like the angle is 312°."

11. Is the student correct? Explain.

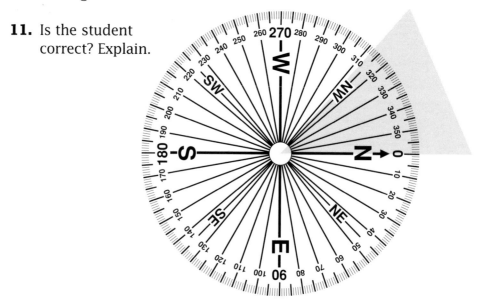

11. No, the student is not correct. When you use a compass card to measure an angle, you always measure to the right. This student measured the angle to the left.

Students can use copies of Transparency Master 4 to make some of the star-like patterns found in the Alhambra. Each section of the transparency looks like the diagram below.

Hints and Comments

Materials copies of Transparency Masters 4, 5, and 6, optional (one of each per student); scissors, optional (one pair per student)

Overview Students read the section Summary. They also explain how to use a compass card correctly to measure static angles.

Planning Problem **11** may be assigned as homework. Invite students to discuss their answers to problem **11** in class. After students complete Section G, you may assign appropriate activities from the Try This! section, located on pages 47–50 of the Student Book, as homework.

Comments about the Problems

11. Homework This problem can be assigned as homework. Estimating an angle before measuring it ensures the correct use of the compass card.

Extension The Alhambra, a citadel and palace in Granada, Spain, is one of the finest examples of Moorish architecture in Spain. The art in the Alhambra contains beautiful geometric patterns.

First, have students make an eight-pointed star using two sections. Then have them make a 12-pointed star using three sections.

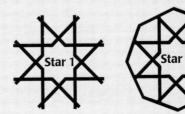

Have students use copies of Transparency Master 5 to make Star 1. Then ask: *Are there any right angles in Star 1?* [Yes, there are 24 right angles.] *What other kinds of angles are in Star 1?* [obtuse angles of 135° and acute angles of 45°]

Have students use copies of Transparency Master 6 to make Star 2. Ask students to compare Star 1 and Star 2 [Star 2 is basically Star 1 with a border around it.] Then have them describe the angles in Star 2. [The angles are 135° and 67.5°.]

Ask students to describe a way to make either Star 1 or Star 2 using turns or headings.

Encourage students to make a 16-pointed Star using copies of Transparency Masters 5 and 6.

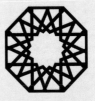

16-pointed star

Work Students Do

The end-of-unit assessment activities include a set of six short assessments and a thematic task, The Floriade Flower Exhibition. The set of activities assesses students' ability to work with angles in various context situations. Students determine headings, create a shape using direction and distance, find a location using headings, and reason about the angles in a geometric shape. The Floriade Flower Exhibition assesses students' ability to plan routes through a flower exhibition. The plan resembles a polar grid. Students give and follow directions from one place in the exhibition to another.

Goals

- indicate a direction (heading) using cardinal directions and degrees

- identify a position using both rectangular and polar grids
- understand and use the relationships between turns and angles
- estimate and measure distances on a map or grid, directions relative to north, and turns and angles
- use the scale on a map to estimate distances
- use directions, headings, turns, and angles to solve simple problems
- understand the relationships among turns, resulting angles, and the number of sides of a regular polygon
- recognize that there are different ways to present information in a map and recognize the consequences of different presentation methods
- understand and use the relationships among directions, headings, turns, and angles
- understand the relationships between the dynamic definition and the static interpretation of angles
- use directions, turns, and angles in combination with scales, distances, and the implicit use of vectors to solve more complex problems

Assessment Opportunities

East Wind Island
I See You and You See Me
The Cutting Machine
Detecting a Fire
Hide-and-Seek

Detecting a Fire
The Floriade Flower Exhibition

The Floriade Flower Exhibition

Hide-and-Seek
Patchwork
The Floriade Flower Exhibition

Hide-and-Seek
The Floriade Flower Exhibition

I See You and You See Me
The Floriade Flower Exhibition

Patchwork

The Floriade Flower Exhibition

The Cutting Machine
Detecting a Fire
The Floriade Flower Exhibition

Patchwork

The Floriade Flower Exhibition

Pacing

- The six short assessment activities and the Floriade Flower Exhibition assessment combined will take approximately two 45-minute class sessions.

About the Mathematics

These end-of-unit activities assess the majority of the goals of the *Figuring All the Angles* unit. Refer to the Goals and Assessment Opportunities chart on the previous page for information regarding the goals that are assessed in each assessment activity. The problems do not ask students to use any specific strategy. They have the option of using their own strategies or any strategy with which they feel comfortable.

Materials

- Assessments, pages 152–161 of the Teacher Guide (one of each per student)
- compass cards from Transparency Master 2, pages 153–155 and 157–161 of the Teacher Guide (one per student)

Planning Assessment

Students should work on these assessment problems individually or in pairs, depending on the nature of your class and your goals for assessment. Make sure that you allow enough time for students to complete the problems. Students are free to solve these problems in their own way.

Scoring

In scoring the assessment problems, the emphasis should be on the strategies used rather than on students' final answers. Since several strategies can be used to answer many of the questions, the strategy a student chooses will indicate his or her level of comprehension of the problem. For example, a concrete strategy supported by drawings may indicate a deeper understanding than an abstract computational answer. Consider how well students' strategies address the problem, as well as how successful students are at applying their strategies in the problem-solving process.

EAST WIND ISLAND

Use additional paper as needed.

Sara lives on Bonaire, an island in the Northern Hemisphere. At noon the sun shines from the south. On Bonaire, the wind is very strong and always blows from east to west. On the right is a picture Sara took of a tree on a cliff on Bonaire. The picture was taken at noon.

1. What direction was Sara facing when she took the picture?

I SEE YOU AND YOU SEE ME

2. Diana and Jeffrey are sitting in a classroom. When Diana looks directly at Jeffrey, she sees him at a heading of 50°. At what heading does Jeffrey look directly at Diana?

1. Answers will vary. Possible answers:

   Sara was facing south. Because the wind is blowing from the east, the branches point west. I made the following diagram and labeled the other directions.

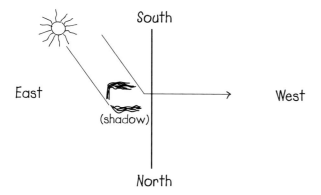

   Since the tree is bending to the west, east is left. Sara sees what I see, so she *must* be looking south.

Materials East Wind Island and I See You and You See Me assessments, page 152 of the Teacher Guide (one per student)

Overview Students indicate a direction using cardinal directions.

Planning Students may work individually or in pairs on these problems.

Comments about the Problems

1. Students can support their reasoning with drawings. The reasonableness of students' explanations is important in assessing their work.

2. To examine the relationship between the two relative headings, encourage students to draw a diagram as shown in the solution column. Students should recall that headings are always measured in a clockwise direction. Jeffrey must make more than a 180° turn to see Diana.

2. The heading is 50° + 180°, or 230°.

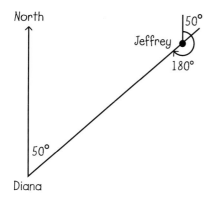

THE CUTTING MACHINE

Use additional paper as needed.

Below is a cutting machine that has instructions for cutting out a shape. The diagram on the right shows where the machine will begin cutting.

1. On the diagram below, draw the figure that will be cut out. You may use your compass card.

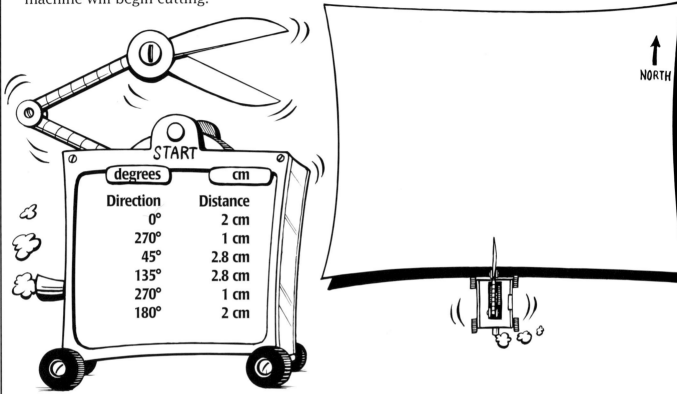

Direction	Distance
0°	2 cm
270°	1 cm
45°	2.8 cm
135°	2.8 cm
270°	1 cm
180°	2 cm

NORTH

2. How would you describe the figure if you had to use turns instead of headings? Use turns to complete the new list of instructions on the right.

Note: The cutting machine always starts in the direction 0 degrees.

3. Think of a figure with six sides (straight lines) and two right angles. Draw the figure and list instructions for cutting it out.

degrees	left/right	cm
TURNS		**DISTANCE**
none	start straight	2 cm
_____°	to the _____	_____ cm
_____°	to the _____	_____ cm
_____°	to the _____	_____ cm
_____°	to the _____	_____ cm
_____°	to the _____	_____ cm

1. Answers will vary. Possible answer:

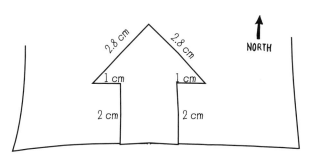

2. Answers will vary. Possible answer:

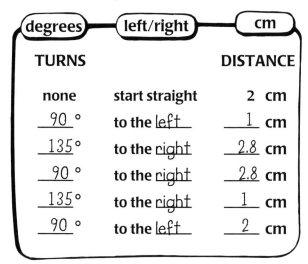

degrees	left/right	cm
TURNS		**DISTANCE**
none	**start straight**	**2 cm**
90 °	to the left	1 cm
135°	to the right	2.8 cm
90 °	to the right	2.8 cm
135°	to the right	1 cm
90 °	to the left	2 cm

3. Answers will vary, but the figure must have six sides and two right angles. Some students may give directions using turns or headings alone; others may use combinations of turns and headings.

Possible figure:

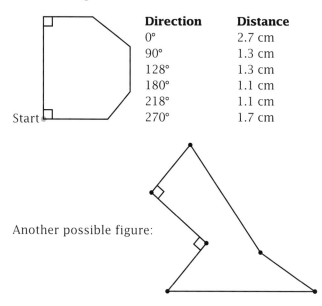

Direction	Distance
0°	2.7 cm
90°	1.3 cm
128°	1.3 cm
180°	1.1 cm
218°	1.1 cm
270°	1.7 cm

Start

Another possible figure:

Materials The Cutting Machine assessment, page 153 of the Teacher Guide (one per student); compass cards (one per student)

Overview Students draw a figure from instructions that use degree headings and describe the figure using turns. They then think of another figure and provide directions for cutting it out.

Planning Students can work individually or in pairs on this assessment activity. You may want to remind students that turns can be either to the left or to the right.

Comments about the Problems

3. Students may interpret "two right angles" to be an inner right angle and an outer right angle, that is, a turn and a resulting angle. As long as a students' answers are consistent, you should give full credit. The six sides of the figure do not have to be the same length.

Use additional paper as needed.

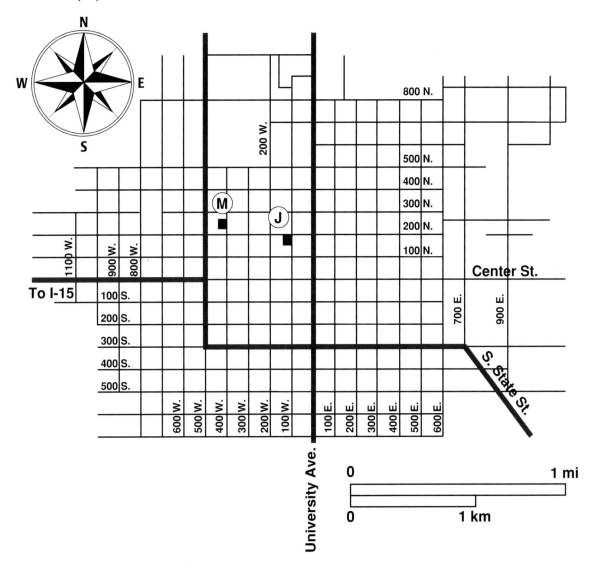

Somewhere in the city pictured above, there is a fire. Maria, who lives on 400 W. (**M** on the map), sees the column of smoke in the northeast. Julio, who lives on 100 W. (**J** on the map), sees the column of smoke directly north from his house.

1. On the map above, show where the fire is.

2. A second fire is reported. Maria and Julio see this new column of smoke in the same direction and in the same heading. Where is the new fire? (There is more than one possibility.) Describe the location(s) from each house using headings and indicate the possible fire location(s) on the map.

1. See map below. The fire is located at the intersection of northeast relative to Maria and north relative to Julio.

2. See map below. Julio and Maria see a fire in the northwest at a heading of around 282° or in the southeast at a heading close to 102°, anywhere along the line MJ, but not between their two homes.

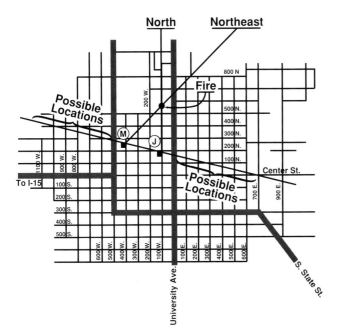

Materials Detecting a Fire assessment, page 154 of the Teacher Guide (one per student); compass cards (one per student)

Overview Students locate two fires (on a rectangular grid) using directions and headings.

Planning Students can work individually or in pairs on this assessment activity.

About the Mathematics This activity assesses students' ability to indicate a direction using degrees, identify a position using both rectangular and polar grids, and use directions and headings in combination with distances to solve a complex problem.

Comments about the Problems

2. Students can be creative in describing the fire location. Answers will vary in degree of precision.

HIDE-AND-SEEK

Use additional paper as needed.

Pete, Hannah, and Josh are playing hide-and-seek in the park. It is Pete's turn to find his friends. He starts at the flagpole (X on the map). First, Pete walks southwest 30 steps. Then he turns and walks 20 steps east. There he finds Josh, who needs to run back to the flagpole before Pete tags him.

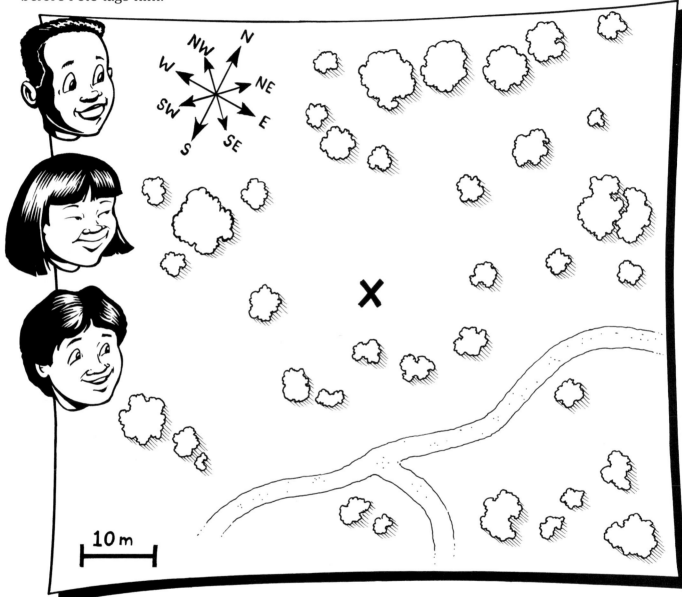

1. Where does Pete find Josh? Show the location on the map.

2. Use your compass card to help you give Josh directions back to the flagpole.

1. Answers will vary. Possible student response:

I decided that five steps are the width of my pencil eraser and measured the distances. I put a big J on the map to mark the spot.

2. Josh should run at a heading of about 5°.

Materials Hide-and-Seek assessment, page 155 of the Teacher Guide (one per student); compass cards (one per student)

Overview Students use wind directions and headings to find a location on a map.

Planning Students can work individually or in pairs on this assessment activity.

About the Mathematics This activity assesses students' ability to indicate a heading using wind directions or degrees, use the scale on a map, and estimate and measure distances on a map and directions relative to north.

Comments about the Problems

2. Where Pete finds Josh will depend on the length the student uses to represent each of Pete's steps. Students should indicate their choice for the length of one (or two or five) step(s). The ratio between the distance Pete walks southwest to the distance he walks east should be 3:2.

PATCHWORK

Use additional paper as needed.

Mitchel has decided to make a quilt like the one on the right. In the center is a star made from six pieces of material. Mitchel used the following pattern to make these six pieces.

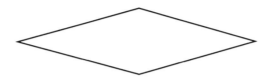

1. Is it possible to make the star with the pieces Mitchel cut? Explain why or why not. You may use a compass card to help you.

It is possible to make many figures using pieces shaped like the one below.

One angle in the figure is labeled 30°.

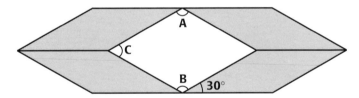

2. Explain, without measuring, how large angles A, B, and C are.

1. No. To make a six-pointed star, you need pieces with 60° angles. The shape given has a 30° angle. Students may explain their answer with a drawing or by arguing that the sum of the measures of the angles at the center of the star must be 360°.

2. Explanations will vary. Sample response:

   Angles A and B are 120°, and angle C is 60°. To find the measure of angle B, subtract 60° (2 × 30°) from 180°, which is 120°. Angle A is the same as angle B.

   To find the measure of angle C:
   I know that the measures of angles A, B, C, and D add up to 360°.

   The measures of angles A and B add up to 240° (120°+120°=240°).

   The measures of angles C and D must add up to 120° (360°−240°=120°).

   Since angle C equals angle D, I know that angle C is 60° (120°÷2=60°).

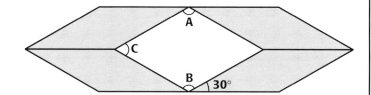

Materials Patchwork assessment, page 156 of the Teacher Guide (one per student)

Overview Given a figure, students investigate whether or not they can make a star with it.

Planning Students can work individually or in pairs on this assessment activity.

About the Mathematics This activity assesses students' ability to estimate and measure angles and understand the relationships between angles and the relationships between the dynamic definition and static interpretation of angles.

Comments about the Problems

2. Students will need to make the following assumptions:

   1) the small angle adjacent to angle B is also 30°;

   2) angles A and B are the same size; and

   3) angles C and D are the same size.

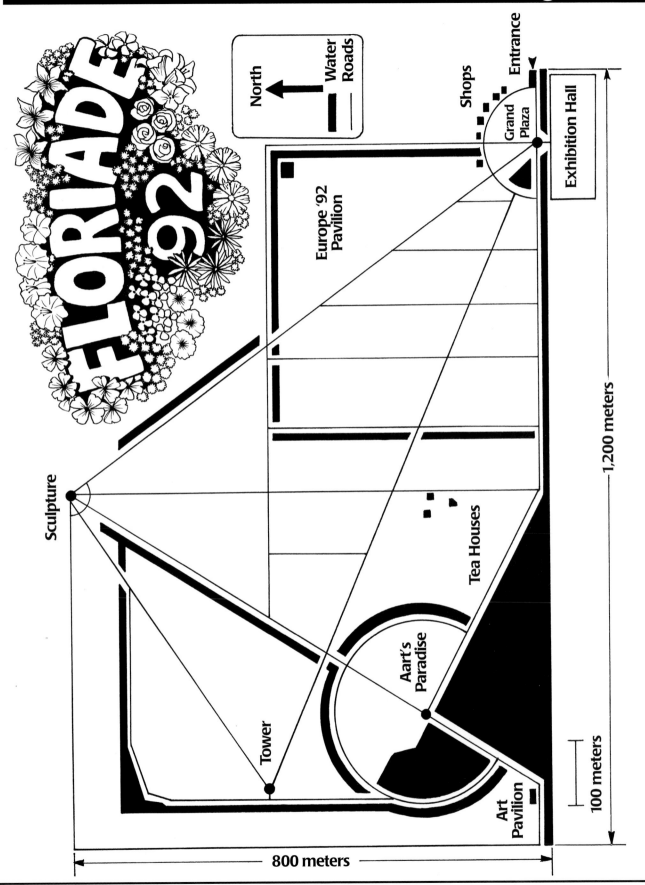

FLORIADE '92

North

Water
Roads

Shops

Entrance

Grand Plaza

Exhibition Hall

Europe '92 Pavilion

Sculpture

Tea Houses

Aart's Paradise

Tower

Art Pavilion

1,200 meters

100 meters

800 meters

Materials The Floriade Flower Exhibition assessment, pages 157–161 of the Teacher Guide (one of each per student)

Overview Students become familiar with the map of the Floriade Flower Exhibition.

Planning You may want to have a class discussion before students begin to work on this assessment. Review the different locations and lengths on the map. Students can work in pairs on these assessment activities.

Use additional paper as needed.

Every 10 years, the European community stages a huge flower show—The Floriade. This event, considered the greatest flower show on Earth, draws thousands of visitors from all around the world. In 1992, the Floriade took place at an exhibition ground designed especially for the event. The grounds shown on the exhibition map measure 800 by 1,200 meters.

Jamaal and Jordan decided to visit the exhibition. Before they entered, they made a note of where the entrance was—in case they got lost.

1. Describe the location of the entrance.

Because Jamaal and Jordan thought they might get separated in the crowd, they agreed on a meeting place far away from the entrance and the main attractions so that they could spot each other.

2. a. Choose a place where Jamaal and Jordan might have decided to meet. Show this on the map.

Jamaal and Jordan decided to go to this place first.

b. In the space below, write directions from the entrance to the place you chose.

1. Descriptions will vary. The entrance is located in the southeast corner of the park. Some students may say only south or east; others may say right or bottom. Neither of these descriptions is complete.

2. **a.** Locations will vary. The location should be away from the entrance and the main attractions so that Jamaal and Jordan can easily spot each other.

   b. Responses will depend on answers given in problem **2a.** Here is one possible set of directions: From the Grand Plaza, walk toward the sculpture (325° heading) until you reach the second street (290 meters).

Materials The Floriade Flower Exhibition assessment, pages 157–161 of the Teacher Guide (one of each per student); compass cards (one per student)

Overview Students describe a location using wind directions. They choose a meeting place on a map and write directions on how to get there.

About the Mathematics This activity provides multiple opportunities to assess students' thinking, reasoning, and understanding of angles, turns, directions, distances, and coordinate systems.

Planning Students can work in pairs on these assessment activities.

Comments about the Problems

2. b. When evaluating student work, consider whether or not the place selected is away from the entrance and the main attractions (you may give extra credit for a centrally located spot) and that the directions are clear.

Use additional paper as needed.

Jordan decided to go to Aart's Paradise, then to the Europe '92 Pavilion, and finally to the Grand Plaza. As he walked, he planned to visit the tower and the sculpture.

Fae entered the exhibition grounds and explained to Jamaal that she was looking for Jordan. Jamaal told her the places Jordan had planned to also visit, but didn't really know where Jordan was.

3. Draw Jordan's path on the exhibition map. Write directions (using turns or headings) that describe the path.

As Jamaal and Fae walked away from the entrance, they saw a large sign.

4. What wind direction describes the location of the restaurant from the entrance?

3. Answers will vary. Students may use headings, directions, turns, and so on. The description of the route should follow the path drawn on the map (see below). The route is designated with arrows. Here is one possible description:

From the meeting place, head south. Then head due west, follow the road to Aart's Paradise. Head (32°) toward the sculpture. Take the first road left toward the tower. From the tower travel northeast to the sculpture. Then head 145° to the first road and turn left. At the Europe '92 Pavilion head south, back to the Grand Plaza.

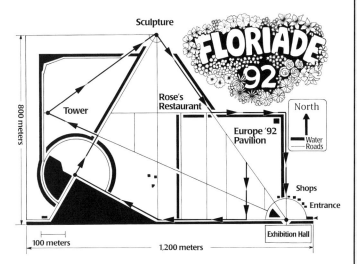

4. northwest

Materials The Floriade Flower Exhibition assessment, pages 157–161 of the Teacher Guide (one of each per student); compass cards (one per student)

Overview Students draw and write directions using turns or headings. Then they describe a location using a wind direction.

Planning Students can work in pairs on this assessment.

Comments about the Problems

3–4. These problems assess students' ability to use directions, headings, turns, and angles to solve simple and complex problems.

Use additional paper as needed.

Jamaal and Fae found Jordan. They all visited the Grand Plaza. Here is a detailed plan of the Grand Plaza.

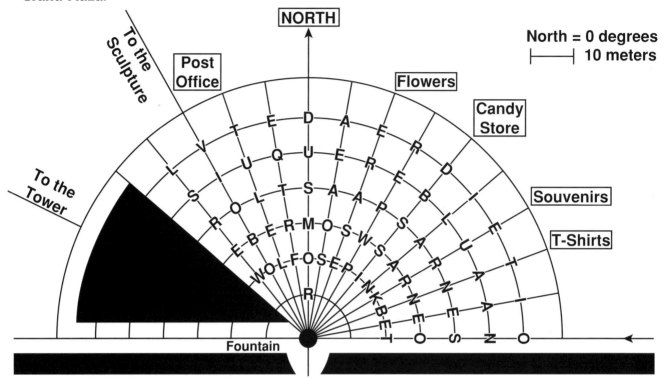

The black dot represents a water fountain. Shops are located around the edge of the plaza. From the fountain to the shops, there is a series of partial circles 10 meters apart painted on the concrete. Fourteen lines, running from the fountain to the edge of the plaza, have also been painted on the concrete. The lines represent headings from the fountain.

5. From the fountain, what is the heading of the candy store?

5. The heading relative to the fountain is 40°.

Materials The Floriade Flower Exhibition assessment, pages 157–161 of the Teacher Guide (one of each per student); compass cards (one per student)

Overview Students find a heading for a location on a polar grid.

Planning You may want to have a class discussion before students begin to work on the rest of the assessment. Make sure students understand that north is at 0° and each line then to the right of north is a turn of 10°. Students can work in pairs on these assessment activities.

Comments about the Problems

5. This problem assesses students' ability to identify a position using a polar grid and to use directions and headings to solve a simple problem.

Use additional paper as needed.

A letter has been painted at each intersection of line and circle to create a puzzle. Those who solve the puzzle win a small bouquet of flowers. But the designer thinks the painters have made several mistakes. Below are the puzzle instructions. Write down *every* letter you pass. Remember that all headings are relative to the fountain.

6. Use the plan and the directions to solve the puzzle. See if you can figure out what the solution should have been.

- Start at the fountain and walk 20 meters north.
- Turn clockwise and follow the circle until you are at a heading of 20°.
- Follow this heading, away from the fountain, for 30 more meters.
- Walk along the 50-meter circle until you are at a heading of 60°.
- Walk straight to the souvenir store.
- Walk on the edge of the plaza counterclockwise, until you reach the road leading to the sculpture.
- Walk toward the fountain for 30 meters.
- Turn left and walk on the circle until you are at a heading of 340°.
- Walk 10 meters toward the fountain.
- Walk directly to the 350° heading that is 40 meters from the fountain.
- Walk clockwise along this circle for 20 more degrees.
- Walk directly to the 20° heading that is 50 meters from the fountain.
- Walk clockwise along this circle until you are at a heading of 30°.
- Walk 10 meters along this line away from the fountain.
- Walk counterclockwise until you are on a heading that is due north.

7. Find another message and give directions to a classmate so that he or she can discover it.

6. RO SE SAR EBLU E VIO L E T SA R E R EAD

"ROSES ARE BLUE VIOLETS ARE READ"

Students should note the misspelling of the color red and the fact that the colors are reversed.

7. Answers will vary. Possible words are: rose, flowers, and so on.

Materials The Floriade Flower Exhibition assessment, pages 157–161 of the Teacher Guide (one of each per student); compass cards (one per student)

Overview Students read a set of instructions in order to solve a puzzle.

Planning Students can work in pairs on this assessment activity.

Comments about the Problems

6. This problem assesses students' ability to identify a position using a polar grid and to use directions, turns, and angles to solve complex problems.

You may want to begin this problem by following the first two directions as a class. Make sure students understand the terms *clockwise and counterclockwise.*

7. It might be fun to do a few messages as a class activity and have the student who wrote each message lead the discussion.

Glossary

acute angle (p. 100) an angle having a measure less than 90°

as-the-crow-flies distance (p. 28) the shortest distance, as if you could fly straight to a given location

circular or **polar grid** (p. 68) a grid that has a central point, from which "roads" go in all directions; the grid system works with two numbers: a heading (in degrees) and a distance (in miles or kilometers)

compass card (p. 54) a circular card with marks indicating the cardinal directions and the 360 degrees of a circle

course (p. 58) the direction of a part of a route

degree (p. 54) a unit of measure for a direction or for an angle

direction (p. 6) a course in which something is moving from one point to another point

east (p. 16) the cardinal direction directly opposite of west

edge (p. 98) a line that borders a shape

heading (p. 56) a direction used to plot courses, measured in degrees to the right from the direction north

headwind (p. 32) a wind blowing in the opposite direction of a moving object

hexagon (p. 90) a polygon with six sides

north (p. 6) the cardinal direction directly opposite of south; the direction along any meridian toward the North Pole of Earth

northeast (p. 18) the cardinal direction between north and east

northwest (p. 20) the cardinal direction between north and west

obtuse angle (p. 102) an angle having a measure between 90° and 180°

parallel (p. 24) extending in the same direction, maintaining equal distance, and never intersecting

pentagon (p. 94) a polygon with five sides

perpendicular (p. 24) intersecting lines that form right angles

rectangular grid (p. 72) a grid with horizontal and vertical lines that intersect; each point of intersection can be identified by east–west and north–south names on the grid

resulting angle (p. 92) the angle formed by a turn in a path

right angle (p. 98) an angle having a measure of 90°

south (p. 6) the cardinal direction directly opposite of north; the direction along any meridian toward the South Pole of Earth

southeast (p. 20) the cardinal direction between south and east

southwest (p. 12) the cardinal direction between south and west

sphere (p. 38) a round, three-dimensional figure

square (p. 94) a polygon with four congruent sides and four right angles

taxicab distance (p. 28) a distance that must be driven using roads, similar to movement on a grid

triangle (p. 90) a polygon with three sides

turn (p. 82) a change from one direction to another

west (p. 16) the cardinal direction directly opposite of east

Blackline
Masters

Dear Family,

Dear Family,

Your child is about to begin the *Mathematics in Context* unit *Figuring All the Angles*. In this unit, students use directions, headings, turns, and angles to solve problems. Rectangular and polar grid systems are introduced and students informally use vectors to describe routes.

You can help your child with this material through a variety of at-home activities. Begin by determining the orientation of your home. Talk about which rooms face what direction. Consider the Sun's path across the sky (and the time of day and year) in relation to the position of your house. What does it mean to say that a room has a northern exposure? What room is best situated for viewing a sunrise? a sunset?

If you own a compass, help your child use it to locate visible landmarks. Help your child create a map of a local park, and plot a travel route through it that your child can follow. While traveling in a car, ask your child to use a map to locate the road or street you are on, the direction north, and the direction in which you are traveling. As your child acquires navigation skills, let him or her help you plan a driving route to a place you have never been or a trip by plane or boat to a faraway country.

We hope these suggestions encourage you to create other math activities for your child.

Sincerely,

The Mathematics in Context Development Team

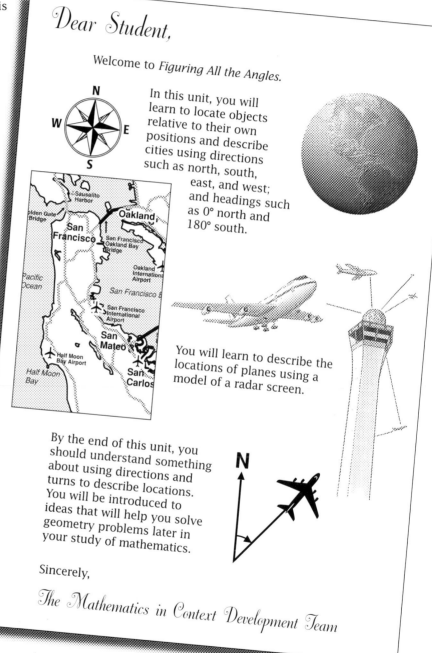

Dear Student,

Welcome to *Figuring All the Angles*.

In this unit, you will learn to locate objects relative to their own positions and describe cities using directions such as north, south, east, and west; and headings such as 0° north and 180° south.

You will learn to describe the locations of planes using a model of a radar screen.

By the end of this unit, you should understand something about using directions and turns to describe locations. You will be introduced to ideas that will help you solve geometry problems later in your study of mathematics.

Sincerely,

The Mathematics in Context Development Team

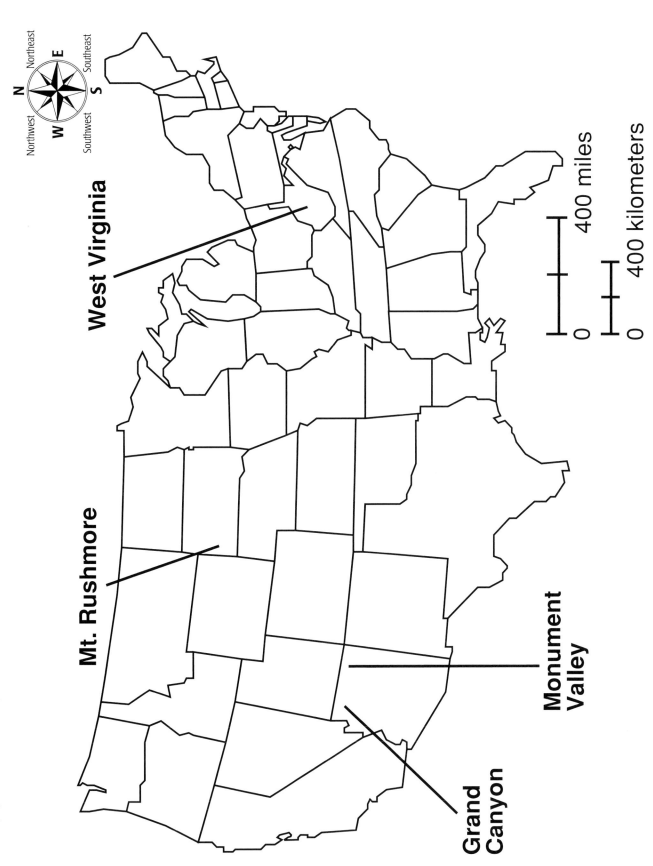

Use with *Figuring All the Angles,* pages 6 and 7.

18. Label the directions of the remaining three wings.

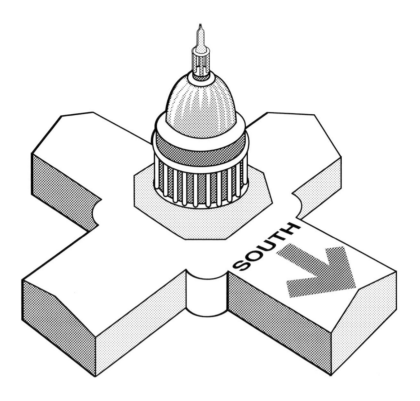

20. Fill in the two missing compass directions.

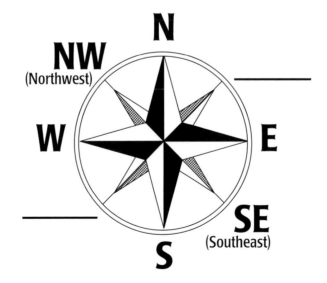

Crystal's house   7th St. N.

Reggie's house

1st St. N.
Town Hall   Main St.

↑ 6th Ave. W.   ↑ 1st Ave. W. ↑ Lincoln Ave.

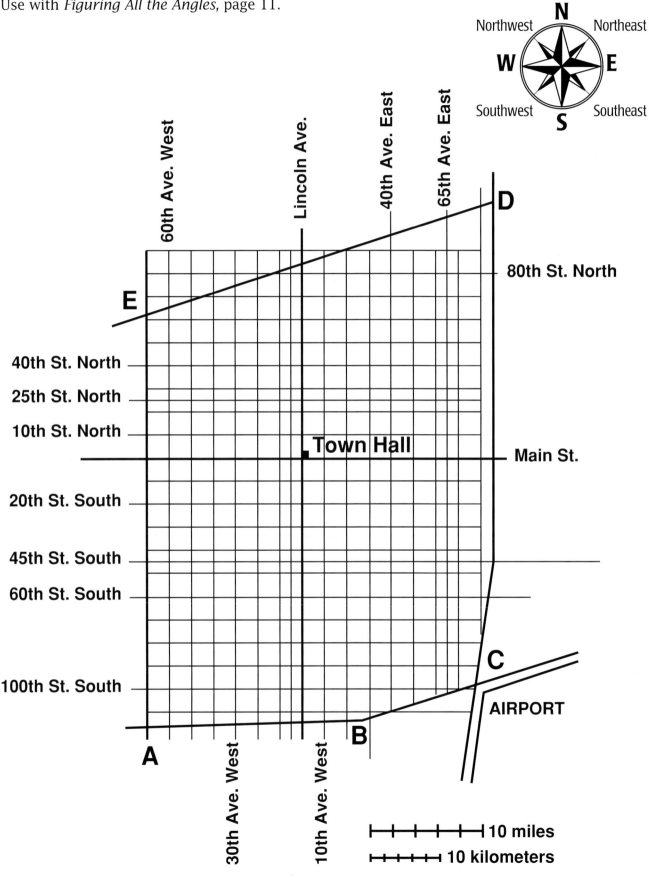

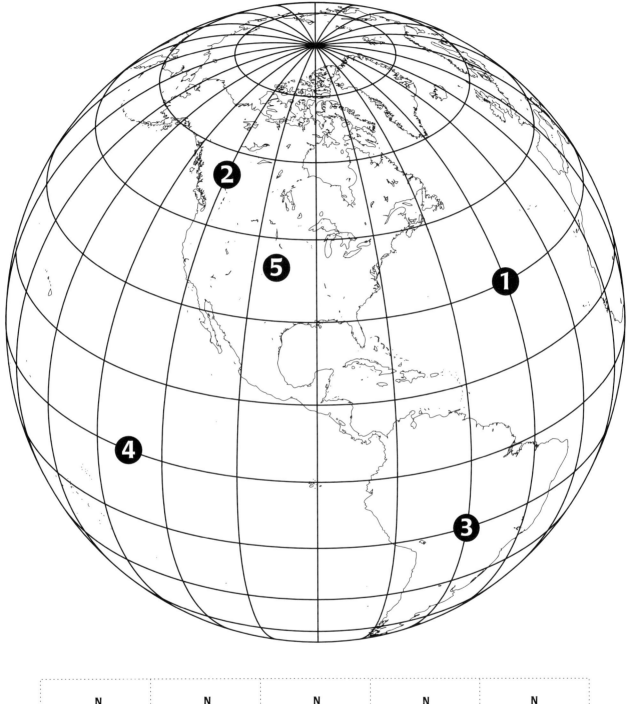

Use with *Figuring All the Angles,* page 21, 23, and 24.

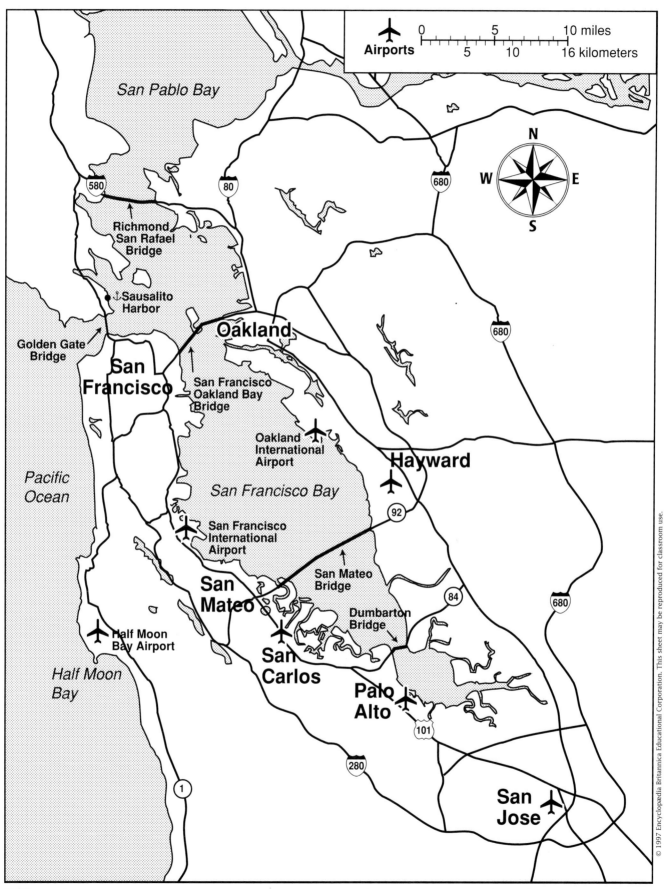

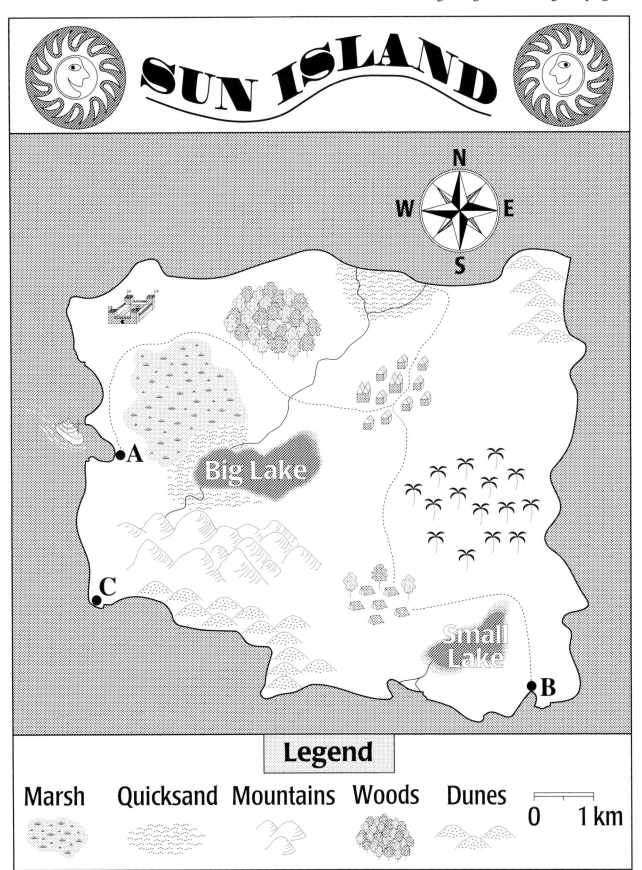

© 1997 Encyclopaedia Britannica Educational Corporation. This sheet may be reproduced for classroom use.

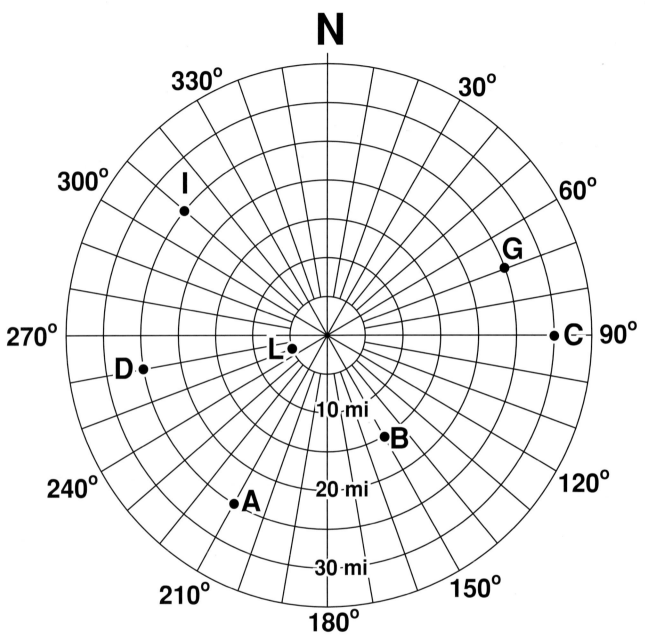

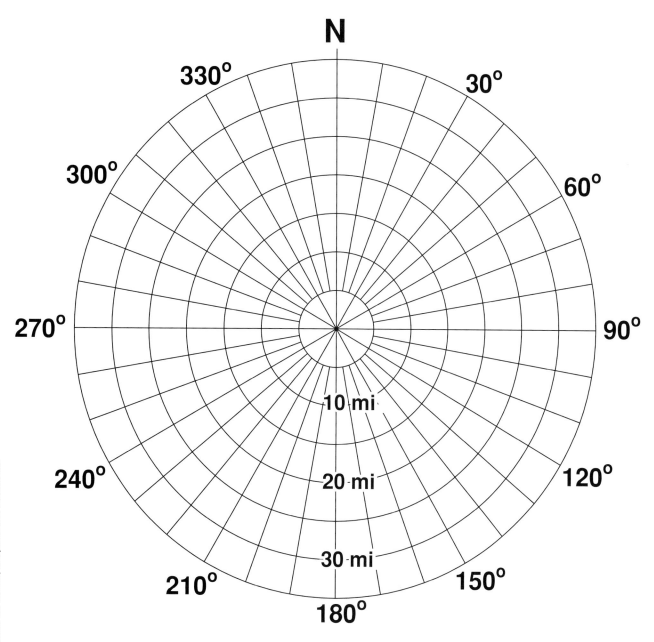

Name _____

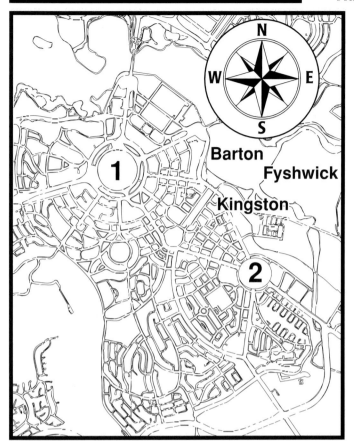

Barton

Fyshwick

Kingston

Use with *Figuring All the Angles,*
pages 30 and 31.

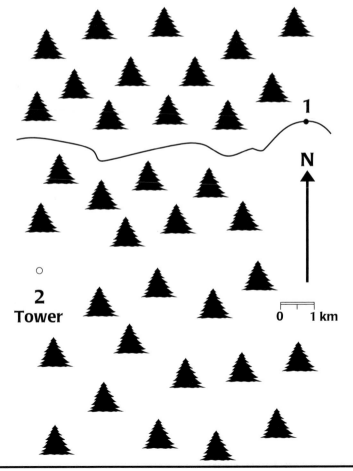

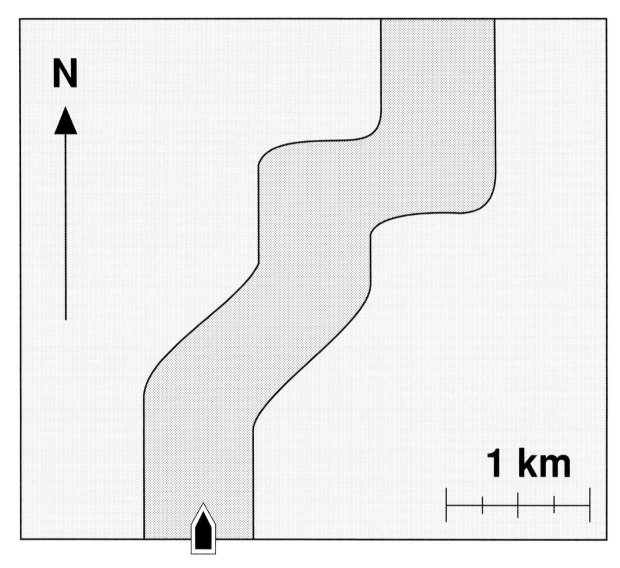

Use with *Figuring All the Angles,* page 22.

Name _____

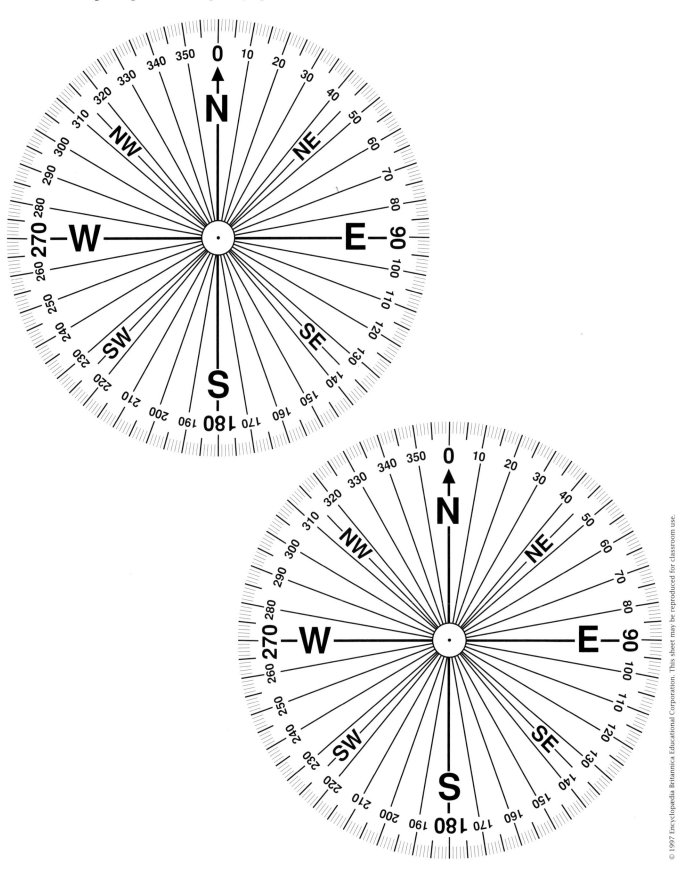

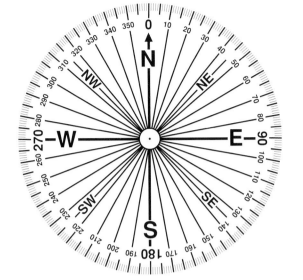

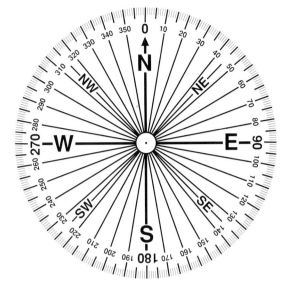

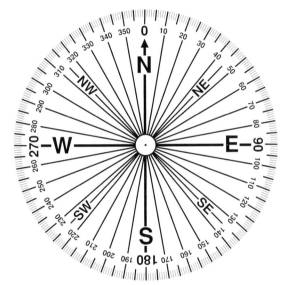

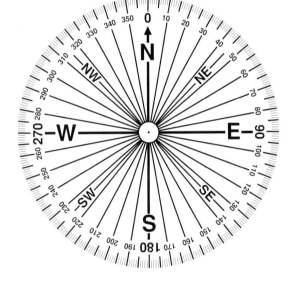

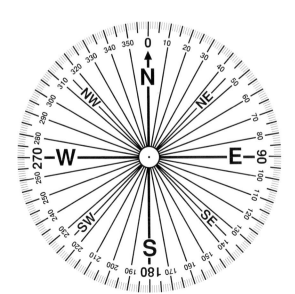

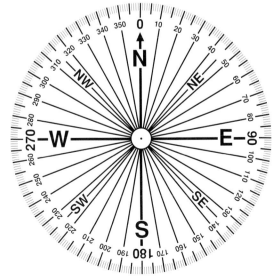

Use with *Figuring All the Angles,* page 34.

Name _____

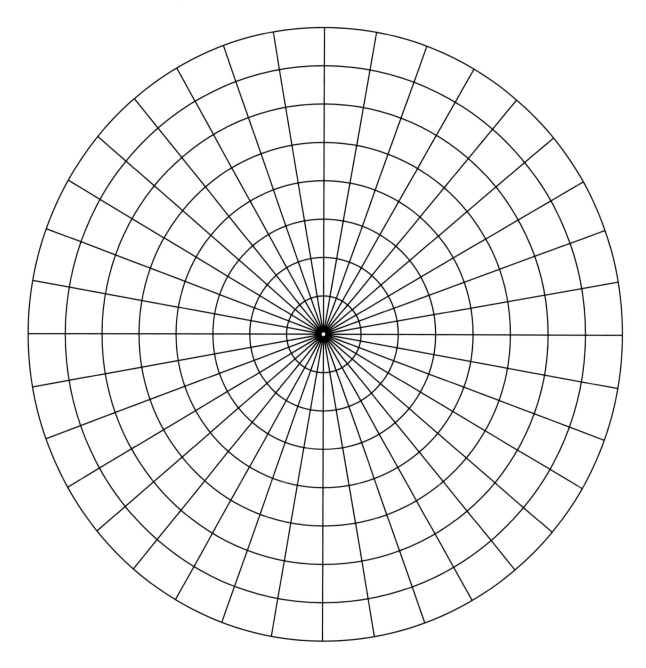

Use with *Figuring All the Angles,* the Alhambra activity, page 109 of the Teacher Guide.

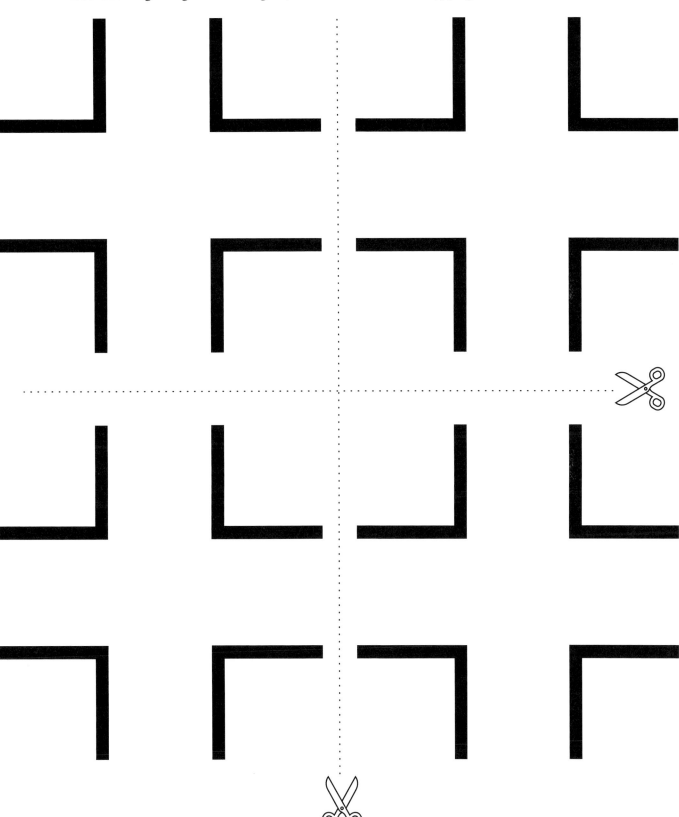

Name _____

Use with *Figuring All the Angles*, the Alhambra activity, page 109 of the Teacher Guide.

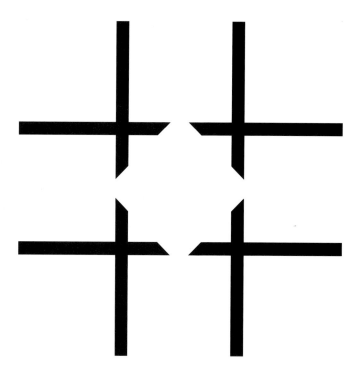

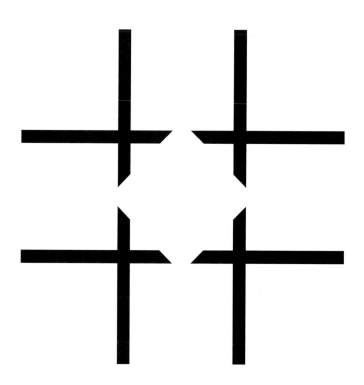

Use with *Figuring All the Angles,* the Alhambra activity, page 109 of the Teacher Guide.

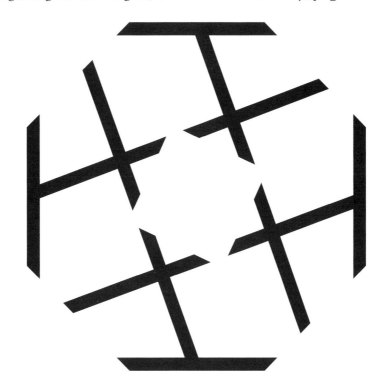

 ···

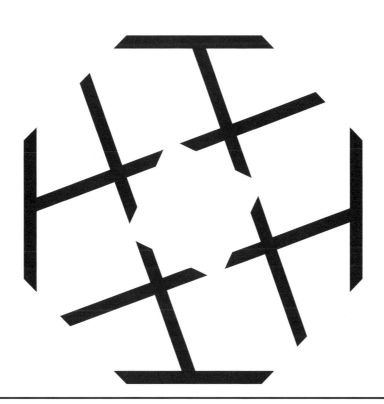

EAST WIND ISLAND

Use additional paper as needed.

Sara lives on Bonaire, an island in the Northern Hemisphere. At noon the sun shines from the south. On Bonaire, the wind is very strong and always blows from east to west. On the right is a picture Sara took of a tree on a cliff on Bonaire. The picture was taken at noon.

1. What direction was Sara facing when she took the picture?

I SEE YOU AND YOU SEE ME

2. Diana and Jeffrey are sitting in a classroom. When Diana looks directly at Jeffrey, she sees him at a heading of 50°. At what heading does Jeffrey look directly at Diana?

Name _____ **Date** _____

THE CUTTING MACHINE

Use additional paper as needed.

Below is a cutting machine that has instructions for cutting out a shape. The diagram on the right shows where the machine will begin cutting.

1. On the diagram below, draw the figure that will be cut out. You may use your compass card.

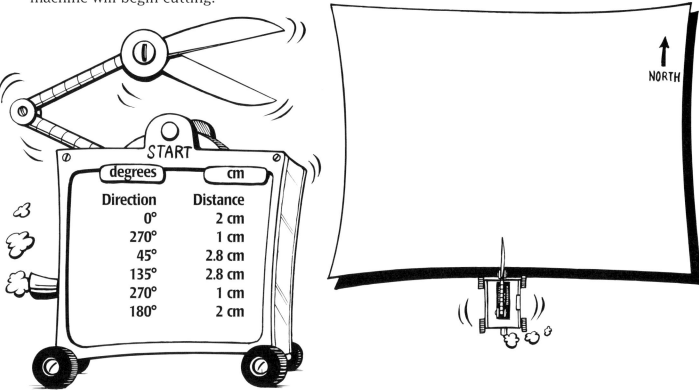

degrees	cm
Direction	Distance
0°	2 cm
270°	1 cm
45°	2.8 cm
135°	2.8 cm
270°	1 cm
180°	2 cm

NORTH

2. How would you describe the figure if you had to use turns instead of headings? Use turns to complete the new list of instructions on the right.

Note: The cutting machine always starts in the direction 0 degrees.

3. Think of a figure with six sides (straight lines) and two right angles. Draw the figure and list instructions for cutting it out.

degrees	left/right	cm
TURNS		DISTANCE
none	start straight	2 cm
_____°	to the _____	_____ cm
_____°	to the _____	_____ cm
_____°	to the _____	_____ cm
_____°	to the _____	_____ cm
_____°	to the _____	_____ cm

© 1997 Encyclopædia Britannica Educational Corporation. This sheet may be reproduced for classroom use.

DETECTING A FIRE

Use additional paper as needed.

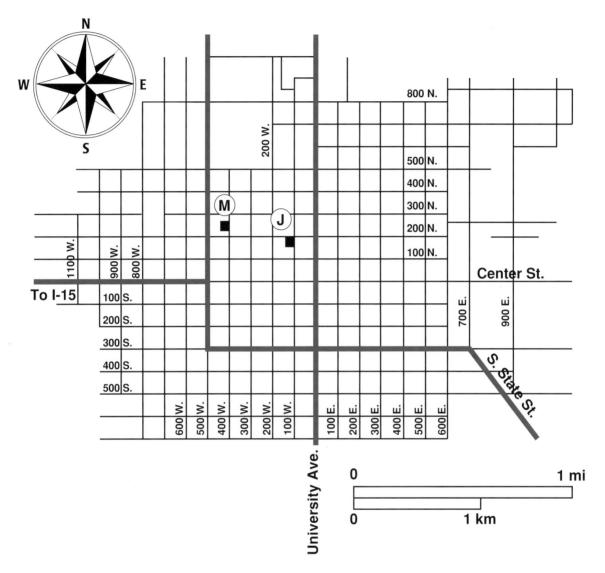

Somewhere in the city pictured above, there is a fire. Maria, who lives on 400 W. (**M** on the map), sees the column of smoke in the northeast. Julio, who lives on 100 W. (**J** on the map), sees the column of smoke directly north from his house.

1. On the map above, show where the fire is.

2. A second fire is reported. Maria and Julio see this new column of smoke in the same direction and in the same heading. Where is the new fire? (There is more than one possibility.) Describe the location(s) from each house using headings and indicate the possible fire location(s) on the map.

HIDE-AND-SEEK

Use additional paper as needed.

Pete, Hannah, and Josh are playing hide-and-seek in the park. It is Pete's turn to find his friends. He starts at the flagpole (X on the map). First, Pete walks southwest 30 steps. Then he turns and walks 20 steps east. There he finds Josh, who needs to run back to the flagpole before Pete tags him.

10 m

1. Where does Pete find Josh? Show the location on the map.

2. Use your compass card to help you give Josh directions back to the flagpole.

PATCHWORK

Use additional paper as needed.

Mitchel has decided to make a quilt like the one on the right. In the center is a star made from six pieces of material. Mitchel used the following pattern to make these six pieces.

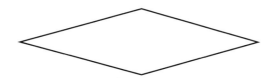

1. Is it possible to make the star with the pieces Mitchel cut? Explain why or why not. You may use a compass card to help you.

It is possible to make many figures using pieces shaped like the one below.

One angle in the figure is labeled 30°.

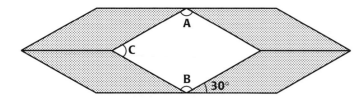

2. Explain, without measuring, how large angles A, B, and C are.

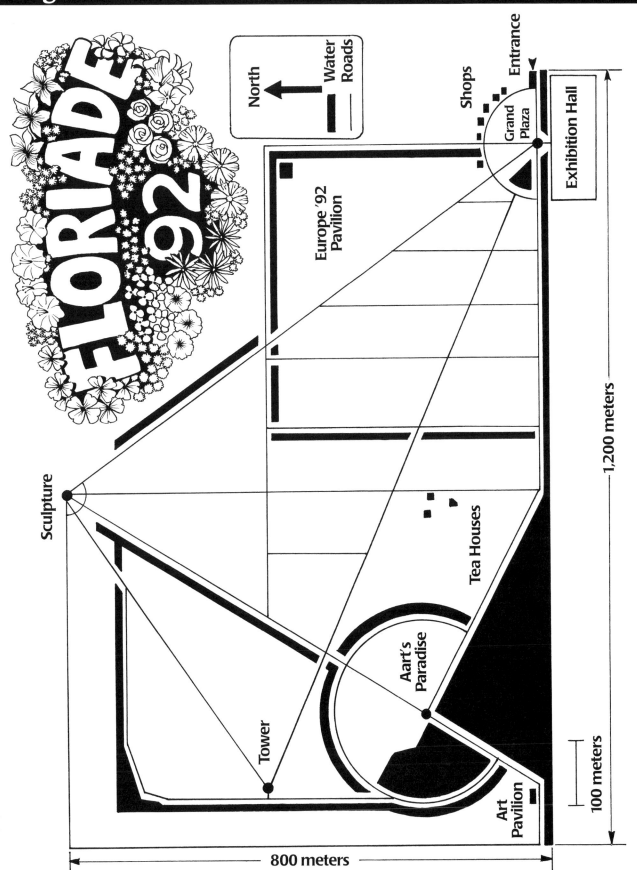

North

Water
Roads

Shops

Entrance

Grand Plaza

Exhibition Hall

Europe '92 Pavilion

1,200 meters

Sculpture

Tea Houses

Aart's Paradise

Tower

Art Pavilion

100 meters

800 meters

Use additional paper as needed.

Every 10 years, the European community stages a huge flower show—The Floriade. This event, considered the greatest flower show on Earth, draws thousands of visitors from all around the world. In 1992, the Floriade took place at an exhibition ground designed especially for the event. The grounds shown on the exhibition map measure 800 by 1,200 meters.

Jamaal and Jordan decided to visit the exhibition. Before they entered, they made a note of where the entrance was—in case they got lost.

1. Describe the location of the entrance.

Because Jamaal and Jordan thought they might get separated in the crowd, they agreed on a meeting place far away from the entrance and the main attractions so that they could spot each other.

2. a. Choose a place where Jamaal and Jordan might have decided to meet. Show this on the map.

Jamaal and Jordan decided to go to this place first.

    b. In the space below, write directions from the entrance to the place you chose.

Name _____ Date _____

Page 3 of 5   **THE FLORIADE FLOWER EXHIBITION**

Use additional paper as needed.

Jordan decided to go to Aart's Paradise, then to the Europe '92 Pavilion, and finally to the Grand Plaza. As he walked, he planned to also visit the tower and the sculpture.

Fae entered the exhibition grounds and explained to Jamaal that she was looking for Jordan. Jamaal told her the places Jordan had planned to visit, but didn't really know where Jordan was.

3. Draw Jordan's path on the exhibition map. Write directions (using turns or headings) that describe the path.

As Jamaal and Fae walked away from the entrance, they saw a large sign.

4. What wind direction describes the location of the restaurant from the entrance?

Use additional paper as needed.

Jamaal and Fae found Jordan. They all visited the Grand Plaza. Here is a detailed plan of the Grand Plaza.

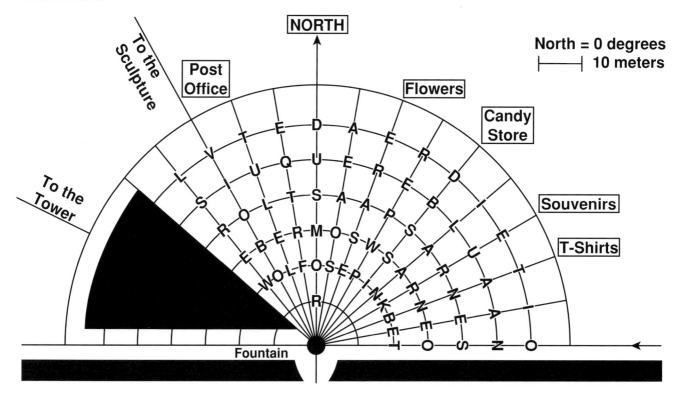

The black dot represents a water fountain. Shops are located around the edge of the plaza. From the fountain to the shops, there is a series of partial circles 10 meters apart painted on the concrete. Fourteen lines, running from the fountain to the edge of the plaza, have also been painted on the concrete. The lines represent headings from the fountain.

5. From the fountain, what is the heading of the candy store?

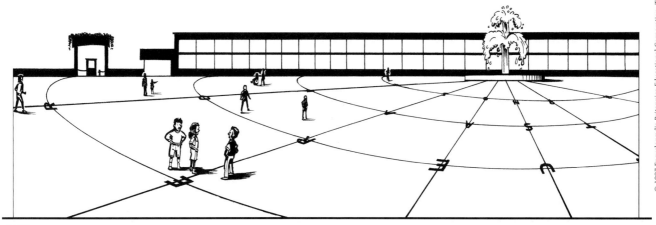

Use additional paper as needed.

A letter has been painted at each intersection of line and circle to create a puzzle. Those who solve the puzzle win a small bouquet of flowers. But the designer thinks the painters have made several mistakes. Below are the puzzle instructions. Write down *every* letter you pass. Remember that all headings are relative to the fountain.

6. Use the plan and the directions to solve the puzzle. See if you can figure out what the solution should have been.

- Start at the fountain and walk 20 meters north.

- Turn clockwise and follow the circle until you are at a heading of 20°.

- Follow this heading, away from the fountain, for 30 more meters.

- Walk along the 50-meter circle until you are at a heading of 60°.

- Walk straight to the souvenir store.

- Walk on the edge of the plaza counterclockwise, until you reach the road leading to the sculpture.

- Walk toward the fountain for 30 meters.

- Turn left and walk on the circle until you are at a heading of 340°.

- Walk 10 meters toward the fountain.

- Walk directly to the 350° heading that is 40 meters from the fountain.

- Walk clockwise along this circle for 20 more degrees.

- Walk directly to the 20° heading that is 50 meters from the fountain.

- Walk clockwise along this circle until you are at a heading of 30°.

- Walk 10 meters along this line away from the fountain.

- Walk counterclockwise until you are on a heading that is due north.

7. Find another message and give directions to a classmate so that he or she can discover it.

Section A. A Sense of Direction

1. Middle East 5; South America 1; South Africa 2; North Korea 4; East China Sea 3

2. because it is in the southern part of Africa

3. northeast

Section B. Finding Your Way

1. northeast–southwest

2. 3rd Ave. East. Explanations will vary. Sample explanation:

Since Madison Street divides the streets into east and west, with all the streets to the right of Madison being east, this means that the third street east of Madison would be 3rd Ave. East.

3. To get to the museum from the train station, go 3 blocks north and then 2 blocks east. To get to Goldstein Park from the museum, go 7 west blocks and then 1 block south.

Section C. Investigating North

1. Answers will vary. Sample answer:

The vertical lines C and D on the graph paper will never meet. Lines A and B on the sphere will meet.

Section D. Directions

1. Walk 4 kilometers at a heading of 120°, turn left 60° and walk 2.5 kilometers, then make a 90° heading and walk 2 kilometers.

Section E. Navigation and Orientation

1. 45 miles

2. 65 miles

Section F. Changing Directions: Turns

1. **a.** 28°

    b. 63°

2. Answers will vary. Sample answer: Make a 35° turn to the left.

Section G. Angles and Shapes

1. **a.** Eight. N and E, NE and SE, E and S, SE and SW, S and W, SW and NW, W and N, and NW and NE

    b. Eight. N and SE, NE and S, E and SW, SE and W, S and NW, SW and N, W and NE, and NW and E

Cover

Design by Ralph Paquet/Encyclopædia Britannica Educational Corporation

Collage by Koorosh Jamalpur/KJ Graphics

Title Page

Illustration by Paul Tucker/Encyclopædia Britannica Educational Corporation

Illustrations

6, 8 Phil Geib/Encyclopædia Britannica Educational Corporation; **16** David Wood/ Encyclopædia Britannica Educational Corporation; **24** Phil Geib/Encyclopædia Britannica Educational Corporation; **42** Paul Tucker/Encyclopædia Britannica Educational Corporation; **44, 54, 58, 66, 70, 72, 82, 86, 92, 106, 112, 114, 118, 120, 122, 124, 126, 128** Phil Geib/Encyclopædia Britannica Educational Corporation

Photographs

10 (top) © Yvonne Navarro; **10 (bottom)** © Paul Buchbinder; **12 (top)** Image obtained from the Corel Professional Photos CD-ROM™; **12 (bottom)** © Ken Griffiths/Tony Stone Images; **14** © Ezz Westphal/ Encyclopædia Britannica Educational Corporation; **16** © Wisconsin Department of Tourism; **52** © Lawrence Manning/Tony Stone Images; **66** © Cameramann International; **80** © Steve Leonard/Tony Stone Images; **88** © Encyclopædia Britannica Educational Corporation; **104** © Elemond Iconografico; **106** © Encyclopædia Britannica Educational Corporation

Mathematics in Context is a registered trademark of Encyclopædia Britannica Educational Corporation. Other trademarks are registered trademarks of their respective owners.